The Hurricane

First published in 1990, this book describes the nature of the hurricane, one of the world's most dangerous weather hazards. It examines the formation, development, movement, and impact of these tropical cyclones, and assesses the ability of science to describe, forecast, and control them.

The Hurricane

Roger A. Pielke

Routledge
Taylor & Francis Group

First published in 1990
by Routledge

This edition first published in 2011 by Routledge
2 Park Square, Milton Park, Abingdon, Oxon, OX14 4RN

Simultaneously published in the USA and Canada
by Routledge
270 Madison Avenue, New York, NY 10016

Routledge is an imprint of the Taylor & Francis Group, an informa business

A Library of Congress record exists under LC Control Number: 89011001

ISBN 13: 978-0-415-61553-2 (hbk)
ISBN 13: 978-0-415-61554-9 (pbk)

The Hurricane

THE HURRICANE

Roger A. Pielke

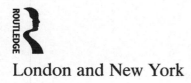

London and New York

First published 1990
by Routledge
11 New Fetter Lane, London EC4P 4EE

Simultaneously published in the USA and Canada
by Routledge
a division of Routledge, Chapman and Hall, Inc.
29 West 35th Street, New York, NY 10001

Typeset by Leaper and Gard Ltd, Bristol
Printed in Great Britain by Richard Clay Ltd,
Bungay, Suffolk

British Library Cataloguing in Publication Data

Pielke, Roger A.
   The hurricane
   1. Hurricane
   I. Title
   551.5′52

   ISBN 0-415-03705-0

Library of Congress Cataloging in Publication Data

Pielke, Roger A.
   The hurricane/Roger A. Pielke.
      p.   cm.
   Includes bibliographical references.
   ISBN 0-415-03705-0
   1. Tropics—Cyclones. 2. Hurricanes—United States.
   I. Title. QC945.P63   1990
   551.55′2—dc20   89-11001

Contents

List of figures   vi
List of tables   ix
Preface   x
Introduction   1

1 Geographic and seasonal distribution   5
  Origin   6
  Movement   8
  *Relation of tropical cyclones to the general circulation
      of the earth*   11
  Tropical cyclone development criteria   25

2 Mechanisms of formation and development   32
  Tropical cyclone formation   32
  Tropical cyclone intensification   34

3 Controls on tropical cyclone movement   48
  External flow   48
  Interaction of the steering current and the hurricane   50
  Internal flow   55

4 Impacts   58
  Ocean impacts   58
  Land impacts   59

5 Tropical cyclone tracks   88
  Tropical cyclone track predictions   89
  Tropical cyclone intensity change predictions   91
  Tropical cyclone-related public forecasts   91
  Seasonal predictions of tropical cyclone activity   96
  Attempts at tropical cyclone modification   97

Appendix A: Atlantic tropical hurricane tracks, 1871–1989   101
Appendix B: Atlantic tropical cyclone track map   220
References   223
Further reading   226
Index   227

Figures

1.1   (a)–(d) Location of first detection of intensifying tropical disturbances, 1952–71.

1.2   (a) Location of first storm origin, 1952–71.
(b) Total number of storms for each genesis region, 1952–71.

1.3   Typical tropical cyclone tracks, 1968–77, in (a) the west Atlantic, (b) the east Pacific, (c) the western north Pacific, (d) the north Indian Ocean, (e) the south Indian Ocean, (f) north and west Australia, (g) the south Pacific.

1.4   (a)–(1) Ten years of tropical cyclone tracks, January–December.

1.5   Probability of at least one tropical storm or hurricane per season.

1.6   Number of tropical storms/hurricanes and hurricanes observed on each day, 1 May to 30 December, 1886–1977.

1.7   Use of the Saffir/Simpson scale to delineate storm intensity.

1.8   Schematic of the general circulation of the earth in the northern hemisphere winter.

1.9   (a)–(d) Seasonal climatology on the depth of the 26°C isotherm.

1.10 Seasonal average winds for 1980–1 at two pressure levels during the periods (a) 1 December to 28 February, (b) 1 March to 31 May, (c) 1 June to 31 August, (d) 1 September to 30 November.

2.1   (a) Development of deep cumulonimbus convection from low-level horizontal wind convergence.
(b) The cumulonimbus region becoming a broader area of deep convection through enhanced low-level wind convergence.

2.2   Schematic of flow (a) towards a low pressure and (b) outwards from a high pressure centre which is deflected to the right by the Coriolis effect.

2.3   A geostationary satellite infra-red image of Hurricane Gloria on 25 September 1985 at 17.00 GMT.

2.4   Schematic of the formation of an eye, as the winds and horizontal pressure gradient force become too strong to permit the air to spiral all the way into the centre of the tropical cyclone.

2.5   Visible satellite image of the hurricane eye in Hurricane Gloria at 17.31 GMT on 24 September 1985.

2.6   Particle trajectories calculated from a numerical model of an asymmetrical hurricane.

2.7 Radar picture of Hurricane Donna taken by the WSR-57 at Key West, 10 September 1960.

2.8 Approximate value of maximum sustained one-minute averaged wind speed in Atlantic hurricanes as a function of central pressure.

2.9 (a) Estimate of maximum potential sustained wind speed and minimum central sea level pressure as a function of sea surface temperature.
(b) Scatter diagram of monthly mean sea surface temperature as related to maximum wind for a sample of north Atlantic tropical cyclones.

2.10 The azimuthal mean structure as approximated by the averaging of the sixteen profiles for Hurricane Anita on 2 September 1977.

2.11 Schematic of differences in wind profiles and temperature over ocean and land owing to a relatively cool and rougher land surface.

2.12 (a) Favourable conditions for tropical cyclogenesis.
(b) Observed locations of tropical cyclogenesis.

3.1 Hurricane Betsy track, 27 August to 12 September.

3.2 Combined radar track of Hurricane Betsy, (a) 6–9 September 1965 (Miami–Key West–Tampa radars), (b) 9–10 September 1965 (New Orleans–Lake Charles radars).

3.3 Satellite visible image of Hurricane Gloria at 17.31 GMT on 25 September 1985, showing a cirrus outflow jet moving south-east from the storm.

3.4 Numerical model simulation of the influence of the island of Taiwan on a hurricane track.

3.5 Daytona Beach radar track of eye of Hurricane Dora, 8–10 September 1964.

4.1 (a)–(gg) Aerial photographs of Eloise (1975), Gladys (1975), and David (1979), in increasing order of flight level wind speeds, ranging from 6.9 to 63.0 ms^{-1}, and estimated one-minute averaged winds at 65 ft (19.8 m), ranging from 14 to 110 kn.

4.2 (a) Observation of the passage of Hurricane Kate, 20 November 1985, as monitored by a floating oceanic buoy at 26°N and 86°W.
(b) Track of Hurricane Kate, 19–22 November 1984.

4.3 Estimated storm surge owing to a level 5 hurricane landfalling south of Miami, Florida.

4.4 Relation between wind speed and kinetic energy. A density of 1 kg m^{-3} was assumed for this example.

4.5 Observed rainfall in inches along the Atlantic coast of the United States from a north-eastward moving tropical cyclone east of the Appalachian Mountains between 7 and 12 August 1928.

4.6 Schematic of a postulated mechanism for tornadogenesis in a hurricane environment.
(a) Tilting of strong vertical shear of the horizontal wind by cumulus convection in one location (ascent) and compensating subsidence adjacent to the cumulus updraft.

(b) Resultant generation of horizontal eddy as vertical shear of the horizontal wind is tilted to some extent into the horizontal plane.

(c) Development of subsequent cumulus convection over the eddy concentrates and speeds up the horizontal circulation until it becomes a tornado.

4.7   (a) Tornado occurrence in the United States with respect to direction of movement of landfalling hurricanes.

(b) The track of the eyes of Hurricanes Carla, Beulah, and Celia, and the area in Texas covered by hurricane-force winds.

4.8   (a) Types of hurricane damage for different degrees of exposure.

(b) Schematic representation of hazard zones A to D in Texas coastal areas.

4.9   Forecast precipitation in inches (a) for 24 hours and (b) for 12 hours, both ending 12.00 GMT, 20 August 1969.

4.10   Observed rainfall in inches associated with the remnants of Hurricane Camille in central Virginia, 19–20 August 1969.

4.11   Rainfall in inches associated with Hurricane Agnes, 18–25 June 1972 over (a) northern Atlantic coast of the United States and (b) southern Atlantic coast of the United States.

4.12   Cumulative rainfall curves in inches for selected locations during Hurricane Agnes, 18—25 June 1972.

5.1   Forecast positions, generated by six computer models, and the official forecast track for Hurricane Frederic, 7 a.m. CDT, Tuesday, 11 September 1979.

5.2   Average error of 24-hour forecast positions, in nautical miles, 1954–82.

5.3   Successive predicted landfall locations for Hurricane Frederic from 1 p.m. CDT, Monday, 10 September to 1 p.m. CDT, Wednesday, 12 September 1979.

5.4   Probabilities of Hurricane Frederic being within 60 nautical miles (110 km) of selected geographical locations at selected periods up to 72 hours from the time of the advisory bulletin.

5.5   Decision tree used to estimate (a) whether a tropical cyclone will develop and (b) whether an existing cyclone will intensify.

5.6   Hypothesized vertical cross-sections through a hurricane eye wall and rain bands before and after seeding. Dynamic growth of seeded clouds in the inner rain bands provides new conduits for conducting mass to the outflow layer and causes decay of the old eye wall.

Tables

4.1   The Beaufort wind scale for tropical cyclones from state-of-sea
        observations at 1,500 ft (457 m), except for Beaufort numbers
        larger than 19 in which case observations are from 700 mb
        (about 10,000 ft; 3.1 km)
4.2   (a)–(e) Relation between wind and wave characteristics
5.1   Atlantic hurricane names for 1990 and 1991
5.2   The prediction and observed occurrences of seasonal tropical
        cyclone activity in the Atlantic Ocean, Gulf of Mexico and
        Caribbean Sea (1984–9)

Preface

The main purpose of this book is to provide the interested, scientifically oriented reader with an overview of tropical cyclones. Only very limited mathematical discussion is presented, with most of the text stressing physical understanding of this important and interesting atmospheric feature.

One difficulty with writing this text was the use of dimensional units. In the United States, even the scientific community will interchange feet and metres, miles and kilometres, as well as knots, statute miles per hour, and metres per second. The convention adopted in this book is to provide dual dimensional references in two of the three units that are applied to tropical cyclone studies (imperial, metric, and nautical units). I have elected to retain the unit of the original source and have provided one of the equivalent dimensional units parenthetically. When original Figures are used which have imperial units, the Figure caption includes the metric or nautical equivalent. In quotations, the original dimensional unit referred to is retained.

Often in original sources from the United States, miles and miles per hour were used without reference to nautical or statute miles. The general convention which I have adopted is to assume nautical units for marine data sources, and statute miles for land-based observations. While the nautical mile is an international unit (based on 1/60 degree of latitude), the statute mile is an American anachronism. The preferred units should be the metric system. Unfortunately, since this text uses a considerable amount of original material from United States' publications of the 1960s and 1950s and earlier, the inclusion of imperial units is unavoidable. For the convenience of the reader, useful dimensional conversion units are:

1 kilometre = 0.545 nautical mile
1 nautical mile = 1.83 kilometres
                = 1.15 statute miles
1 statute mile = 1.59 kilometres
1 metre = 39.37 inches
        = 3.28 feet
1 foot = 0.3048 metres
1 inch = 2.54 centimetres
       = 25.4 millimetres

1 centimetre = 0.3937 inch
1 knot = 1 nautical mile per hour
       = 1.15 statute miles per hour
       = 0.51 metres per second
1 statute mile per hour = 0.87 knots
                   = 0.44 metres per second

The provision of source material for the book by Doctors Richard Anthes, Morris Bender, Peter Black, Gordon Dunn, Richard Johnson, Miles Lawrence, Robert Merrill, Robert Simpson, Jim Trout, Glenn White, and Hugh Willoughby, and by Professor William Gray is very gratefully acknowledged. Other individuals who have provided valuable perspectives to me on tropical cyclones include Professor Russell Elsberry, Professor William Frank, Mr Paul Hebert, Mr Brian Jarvinen, Mr Gil Clarke, Mr Charlie Neumann, Dr Bob Burpee, Dr John Hope, Dr Herb Riehl, Dr Banner Miller, Dr Greg Holland, Dr Neil Frank, Dr Robert Sheets, and, of course, Dr Joanne Simpson.

I also want to acknowledge the encouragement and the insight into tropical cyclones that I received from Robert Simpson, beginning during my employment in Miami in the early 1970s and continuing during my tenure at the University of Virginia. Bill Gray is thanked for the stimulating discussions regarding tropical cyclones which we have had since my joining the faculty at Colorado State University.

The drafting was completed by Mrs Judy Sorbie-Dunn. Mrs Dallas McDonald provided her standard effective editorial assistance in the completion of this work. The editorial supervision of Ms Emma Waghorn of Routledge is also gratefully acknowledged.

The patience and encouragement of my family, Gloria, Roger Jr, and Tara during the completion of this work is greatly appreciated.

The author welcomes comments regarding the text.

Roger A. Pielke

Introduction

On the evening of 17 August 1969, the residents of coastal Mississippi received the following bulletin from the National Hurricane Center in south Florida:

BULLETIN 9 PM CDT SUNDAY 17 AUGUST 1969

. . . CAMILLE . . . EXTREMELY DANGEROUS . . . CENTER HAS PASSED MOUTH OF THE MISSISSIPPI RIVER . . . CONTINUES TOWARD THE MISSISSIPPI ALABAMA COAST . . .

HURRICANE WARNINGS ARE IN EFFECT FROM NEW ORLEANS AND GRAND ISLE LOUISIANA EASTWARD ACROSS THE MISSISSIPPI . . . ALABAMA . . . AND NORTHWEST FLORIDA COAST TO APALACHICOLA. GALE WARNINGS ARE IN EFFECT FROM MORGAN CITY TO GRAND ISLE. CONTINUE ALL PRECAUTIONS.

WINDS ARE INCREASING AND TIDES ARE RISING ALONG THE NORTHERN GULF COAST FROM GRAND ISLE EASTWARD. HURRICANE FORCE WINDS ARE NOW OCCURRING OVER EXTREME SOUTHEAST LOUISIANA AND WILL BE SPREADING OVER MOST OF THE WARNING AREA WITHIN THE NEXT FEW HOURS.

THE FOLLOWING TIDES ARE EXPECTED TONIGHT AS CAMILLE MOVES INLAND . . . MISSISSIPPI COAST GULFPORT TO PASCAGOULA 15 TO 20 FEET . . . PASCAGOULA TO MOBILE 10 TO 15 FEET . . . EAST OF MOBILE TO PENSACOLA 6 TO 10 FEET. ELSEWHERE IN THE AREA OF HURRICANE WARNING EAST OF THE MISSISSIPPI RIVER 5 TO 8 FEET. IMMEDIATE EVACUATION OF AREAS THAT WILL BE AFFECTED BY THESE HIGH TIDES IS URGENTLY ADVISED.

THE CENTER OF CAMILLE IS EXPECTED TO MOVE INLAND ON THE MISSISSIPPI COAST NEAR GULFPORT BEFORE MIDNIGHT.

SEVERAL TORNADOES ARE LIKELY TONIGHT WITHIN 100

MILES OF THE COAST FROM EXTREME SOUTHEASTERN LOUISIANA TO FORT WALTON BEACH FLORIDA.

HEAVY RAINS WITH LOCAL AMOUNTS 8 TO 10 INCHES WILL SPREAD INTO SOUTHEAST MISSISSIPPI . . . SOUTHWEST ALABAMA . . . AND THE FLORIDA PANHANDLE TONIGHT. ANY FLOOD STATEMENTS NEEDED WILL BE ISSUED BY THE LOCAL WEATHER BUREAU OFFICES.

AT 9 PM CDT . . . THE CENTER OF HURRICANE CAMILLE WAS LOCATED BY NEW ORLEANS AND OTHER LAND BASED RADARS NEAR LATITUDE 29.9 NORTH . . . LONGI-TUDE 89.1 WEST . . . OR ABOUT 35 MILES SOUTH OF GULF-PORT MISSISSIPPI AND 60 MILES EAST OF NEW ORLEANS. CAMILLE WILL CONTINUE NORTHWARD ABOUT 15 MPH.

HIGHEST WINDS ARE ESTIMATED 190 MPH NEAR THE CENTER. HURRICANE FORCE WINDS EXTEND OUTWARD 60 MILES AND GALES EXTEND OUTWARD 180 MILES FROM THE CENTER. THE AIR FORCE RECON FLIGHT INTO CAMILLE THIS AFTERNOON REPORTED A CENTRAL PRESSURE OF 26.61 INCHES.

THOSE IN THE PATH OF THE EYE ARE REMINDED THAT THE WINDS WILL DIE DOWN SUDDENLY IF THE EYE PASSES OVER YOUR AREA BUT THE WINDS WILL INCREASE AGAIN RAPIDLY AND FROM THE OPPOSITE DIRECTION AS THE EYE MOVES AWAY. THE LULL WITH CAMILLE WILL PROBABLY LAST FROM A FEW MINUTES TO ONE HALF HOUR AND PERSONS SHOULD NOT VENTURE FAR FROM SAFE SHELTER.

WINDS GUSTED TO SLIGHTLY OVER 100 MPH AT BOOTH-VILLE LOUISIANA ABOUT 7 PM. NEW ORLEANS WEATHER BUREAU OFFICE WAS REPORTING WINDS 45 TO 50 MPH WITH GUSTS TO NEAR 70 MPH AT 8 PM.

REPEATING THE 9 PM POSITION . . . 29.9 NORTH . . . 89.1 WEST.

THE NEXT ADVISORY WILL BE ISSUED BY THE NEW ORLEANS WEATHER BUREAU AT 11 PM AND BULLETINS AT 1 AND 3 AM CDT.

Over 139 people died that night in Mississippi and south-eastern Louisiana as a direct result of Hurricane Camille.

Two days later, the remnants of what was once Hurricane Camille began to pass over central Virginia. No weather warning nor watch was issued. By morning, 109 people were dead, most of them in Nelson County. Within about six hours, up to 30 in. (762 mm) of rain had fallen in that area, result-ing in the liquidification of soils on mountain slopes in the watersheds of the Tye and Rockfish Rivers. (It is said that the rain was so heavy that birds, which have nostrils on the tops of their beaks, drowned while perched on trees.) This muddy mass moved downhill, burying homes and gorging the rivers with mud and debris from the denuded slopes. It was not until late the following morning, however, when employees did not report to work in Charlottesville from their homes in Nelson County, that the occurrence and magnitude of the flood disaster became known to the outside world.

On 20 August, the remnants of Camille regained tropical storm strength, with winds of 56–60 kn., as it moved rapidly into the Atlantic Ocean east of Virginia. For at least two days, until it finally weakened south of Newfoundland, Tropical Storm Camille posed a threat to shipping.

Losses owing to Camille were estimated in terms of 1969 dollars at $1,420,700,000 (Simpson and Riehl 1981).

This hurricane illustrates the major hazard that such storms present to the United States. Excessive rainfall, particularly in hilly and mountainous areas, can cause serious river flooding and terrain erosion. Coastal regions are exposed to massive damage by hurricane-force winds and tornadoes, ocean inundation, and the rainfall-flooding that is associated with these storms. On the open ocean, the safety of all vessels is threatened by the chaotic seas, often reaching higher than 12 m, that are associated with hurricanes.

This book provides an overview of this natural weather hazard. Chapter 1 presents the observed geographic and seasonal distribution of hurricanes around the world. The physical mechanisms associated with their formation and development is discussed in Chapter 2, while the large-scale and internal controls on their movement are described in Chapter 3. The impact of these storms on the environment is discussed in Chapter 4. Chapter 5 provides an overview of the procedure applied by the National Hurricane Center to forecast the intensity changes in and movement of these storms.

In this book, the following definitions are used:

Tropical low A surface low pressure system which decreases in intensity with height, often becoming a relative high pressure region in the upper troposphere. The centre of a tropical low is warmer than its surroundings.

Tropical cyclone A tropical low with sustained near-surface winds of 27 kn. (31 statute m.p.h.) or greater.

Tropical disturbance An area of enhanced cumulonimbus activity but without a well-defined closed surface wind circulation.

Tropical depression A tropical low with a closed surface wind circulation with speeds less than 27 kn. (31 statute m.p.h.).

Tropical storm A tropical cyclone with winds stronger than 27 kn. (31 statute m.p.h.) but less than 64 kn. (74 statute m.p.h.).

Hurricane A tropical cyclone whose wind speeds are 64 kn. (74 statute m.p.h.) or greater. At this speed and higher, an eye typically occurs. A hurricane is called a typhoon in the western Pacific. (The word typhoon originates from the Chinese word *táifēng*.)

Five levels of hurricanes have been defined. The scale used to catalogue hurricane intensity is referred to as the Saffir/Simpson Damage-Potential Scale.

Hurricane	Central pressure (Inches of mercury in a barometer are given in parentheses)	Maximum sustained winds
Level 1	> 980 mb (28.94 in.)	64–83 kn. (74–95 statute m.p.h.)
Level 2	965–979 mb (28.50–28.94 in.)	84–95 kn. (96–109 statute m.p.h.)
Level 3	945–964 mb (27.91–28.49 in.)	96–113 kn. (110–130 statute m.p.h.)
Level 4	920–944 mb (27.17–27.90 in.)	114–135 kn. (131–155 statute m.p.h.)
Level 5	< 920 mb (< 27.17 in.)	> 135 kn. (> 155 statute m.p.h.)

1 Geographic and seasonal distribution

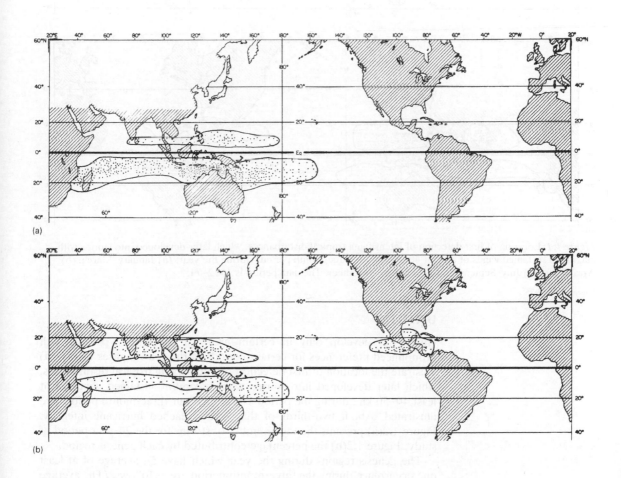

(a)

(b)

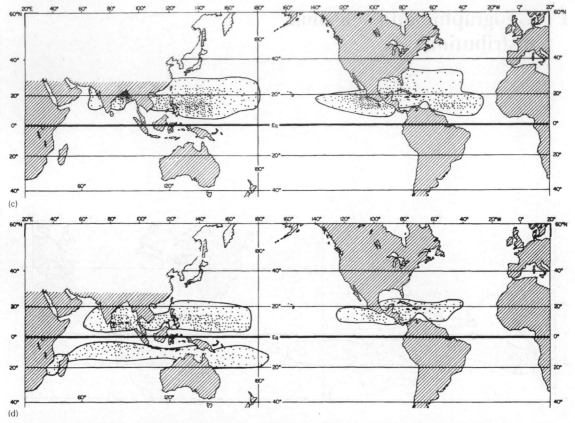

Figure 1.1 Location of first detection of intensifying tropical disturbances, which later developed into storms with maximum sustained winds of at least 40–50 kn. for three-month periods during the year: (a) January–March, (b) April–June, (c) July–September, (d) October–November. The data period is 1952–71.

Source: Gray 1975

ORIGIN

Hurricanes develop only in certain areas of the earth's oceans, with pronounced preferences for certain periods of the year. Figures 1.1(a)–(d) illustrate the location of first detection of intensifying tropical disturbances, which later developed into storms with maximum sustained winds of at least 40–50 kn., during the period 1952–71. Four three-month periods are illustrated. About two-thirds of the storms reached hurricane intensity. Figure 1.2(a) presents the total number of storms for the 20-year period of study; Figure 1.2(b) the percentage contributed by each genesis region.

The genesis regions during the year which have an average of at least one occurrence during the three-month period are as follows. (The average number of occurrences is given in parantheses; data from Gray 1975.)

Jan–Feb–March: South Indian Ocean (8.5)
South Pacific (6.3)

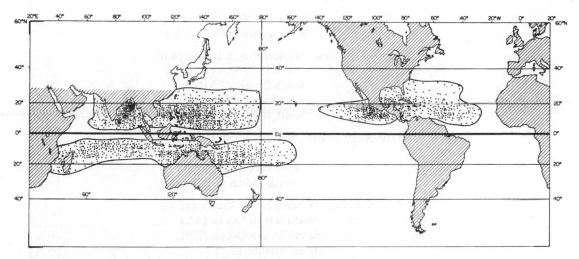

Figure 1.2(a) Location of first storm origin for the 20 years of data (1952–71) for the entire year: a summation of Figures 1.1 (a)–(d)

Source: Gray 1975

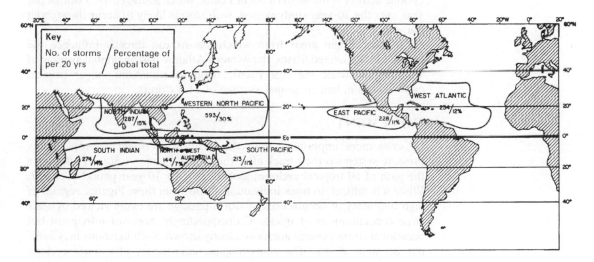

Figure 1.2(b) Total number of storms for each genesis region for the period 1952–71 and the percentage contribution within each region to the global total

Source: Gray 1975

North and West Australia (5.8)
Western North Pacific (1.4)

April–May–June: Western North Pacific (4.0)
North Indian Ocean (3.3)

South Pacific (2.3)
East Pacific (1.9)
South Indian Ocean (1.2)

July–August–Sept: Western North Pacific (11.7)
Western Atlantic (8.4)
East Pacific (7.7)
North Indian Ocean (5.8)

Oct–Nov–Dec: Western North Pacific (12.4)
North Indian Ocean (4.9)
South Indian Ocean (3.9)
West Atlantic (2.7)
South Pacific (2.0)
East Pacific (1.8)

As is evident from these numbers, the most active region of tropical cyclone activity is the western north Pacific which averaged 18.5 storms per year over the 20-year study period, with storms likely to occur throughout the year.

Of those origin areas from which storms can directly influence the weather in the United States, the western Atlantic had an annual average of 11.1 storms, while the east Pacific, with its somewhat longer tropical cyclone season, had an annual average of 11.4 storms.

MOVEMENT

Of even more importance than the location of origin of tropical low pressure systems is their track after they form. Figures 1.3(a)–(g) illustrate the path of all tropical cyclones globally for the 10-year period 1968–77. While it is difficult to track individual storms from these Figures, regions of high frequency of occurrences of storm passage are easily viewed by their large concentrations of tracks. Correspondingly, areas of infrequent but occasional storm passage are more clearly shown. Such locations may actually be more at risk. The climatological infrequency of tropical cyclone occurrences at those sites would tend to provide an aura of insulation from this type of atmospheric hazard.

Tropical cyclone tracks by month for the 10-year period given in Figures 1.3(a)–(g) are illustrated in Figures 1.4(a)–(l). The major changes in storm activity during the year are clearly illustrated in the Figures.

In the west Atlantic, Gulf of Mexico and Caribbean Sea, the probability of at least one tropical cyclone per year entering a 2.5 degree latitude-longitude box is shown in Figure 1.5. The highest probabilities are in the lesser Antilles, through the Yucatan straits between the Gulf and the Caribbean, and north of the Bahamas east of the south-eastern coast of the

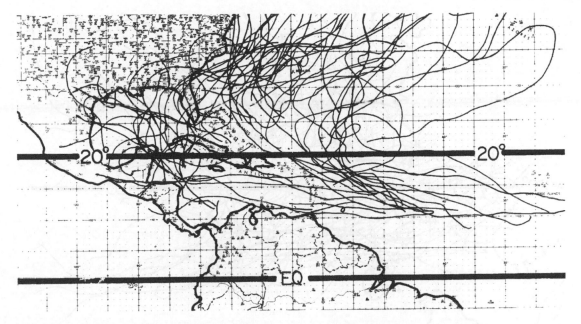

Figure 1.3(a) Typical tropical cyclone tracks in the west Atlantic, for the 10-year period of 1968–77

Source: Gray 1975

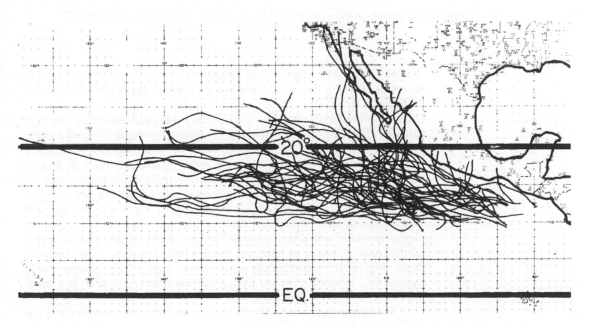

Figure 1.3(b) Typical tropical cyclone tracks in the east Pacific, for the 10-year period of 1968–77

Source: Gray 1975

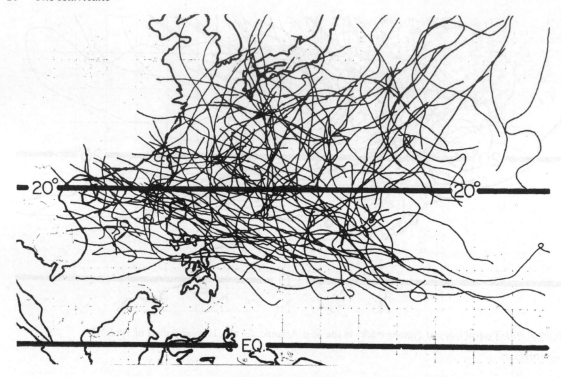

Figure 1.3(c) Typical tropical cyclone tracks in the western north Pacific, for the 10-year period of 1968–77

Source: Gray 1975

United States. The frequency of occurrence of tropical storms and hurricanes anywhere in this ocean basin as a function of day of the year is shown in Figure 1.6.

During 1985 in the Atlantic, over 3,000–4,000 million dollars' worth of damage afflicted property within the United States alone. Tropical cyclone tracks in the Atlantic, Gulf of Mexico and Caribbean region for the period 1900–89 are given in Appendix A, in order to illustrate the year-to-year variability and the trends over time in storm activity in this ocean basin.

In reality, of course, tropical low pressure systems are not points or line segments, as displayed in Appendix A and in Figures 1.3(a)–(g) and 1.4(a)–(l), but have finite areas. A more informative method of presenting this information would be to display the area of gale winds, and the region of different intensities of hurricane force winds, using the Saffir/Simpson scale. Figure 1.7 schematically shows how such a presentation would appear for a storm that evolves from a tropical storm (at time 0) to a hurricane level 2 (at time 36 hrs).

It is important to note two major observed characteristics of tropical cyclones as displayed in Figure 1.7.

- The area of damaging winds extends well beyond the point location of

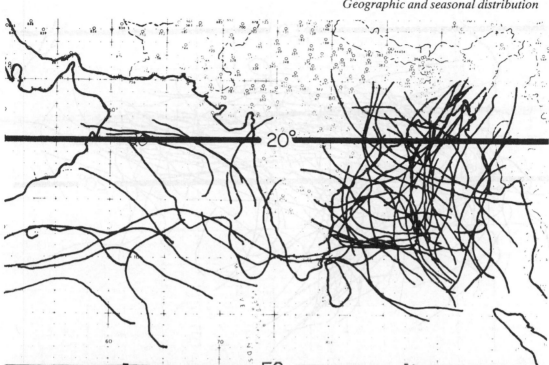

Figure 1.3(d) Typical tropical cyclone tracks in the north Indian Ocean, for the 10-year period of 1968–77

Source: Gray 1975

    hurricane position that is displayed in Figures 1.3(a)–(g) and 1.4(a)–(l) and Appendix A.

- the area of most destruction is concentrated close to the eye of the storm.

RELATION OF TROPICAL CYCLONES TO THE GENERAL CIRCULATION OF THE EARTH

In order to understand why tropical cyclones form where and when they do, an understanding of the general circulation of the earth is required.

    The primary driving force on the earth's atmosphere is the amount and distribution of solar radiation which impinges on the planet. The orbit of the earth around the sun is an ellipse, with an apogee (closest approach) of 1.47×10^8 km (8.0×10^7 nautical miles) in early January and a perigee (furthest distance) of 1.52×10^8 km (8.2×10^7 nautical miles) in early July. The time between the *autumnal equinox* and the following *vernal equinox* in the northern hemisphere (about 22 September – about 21 March) is approximately one week shorter than the remainder of the year, as a result of the earth's elliptical orbit, resulting in shorter winters in the northern hemisphere than south of the equator.

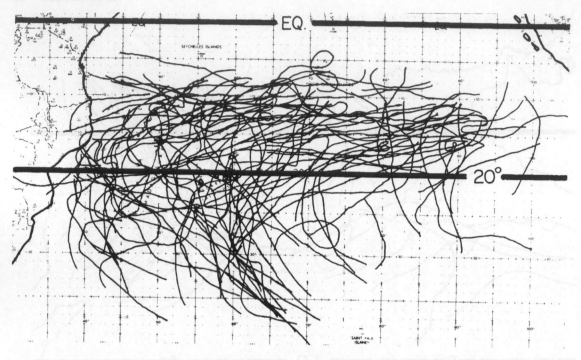

Figure 1.3(e) Typical tropical cyclone tracks in the south Indian Ocean, for the 10-year period of 1968–77

Source: Gray 1975

The earth rotates every 24 hours around an axis that is tilted at an angle of 23.5° with respect to the plane of its orbit. As a result of this tilt, during the summer season in either the northern or southern hemisphere, sunshine is more direct on a flat surface at a given latitude than it is during the winter season. Poleward of 66.5° of latitude, the tilt of the earth is such that, for at least one complete day (at 66.5°) and for as long as six months (at 90°), the sun is above the horizon during the summer season and below the horizon during the winter.

The *troposphere* is the layer of the atmosphere in which most weather occurs. It is characterized by temperatures which generally decrease with height. Its height varies from less than 10 km over polar regions to up to 18 km in the tropics. Above this height, temperatures become nearly constant and then, higher up, increase with height as a result of the absorption of ultraviolet solar radiation during ozone formation. This layer of the atmosphere above the troposphere is called the *stratosphere*. The level at which the temperature decrease with height stops is called the *tropopause*.

As a result of the asymmetric distribution of solar heating described above, during the winter season, high latitudes become very cold in the troposphere because of the long nights. In the summer at high latitudes, the troposphere warms significantly as a result of the long hours of daylight.

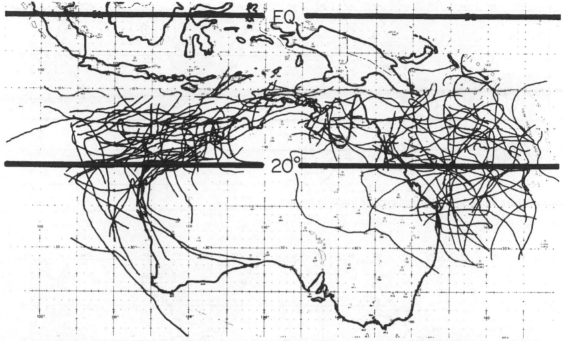

Figure 1.3(f) Typical tropical cyclone tracks in north and west Australian region, for the 10-year period of 1968–77

Source: Gray 1975

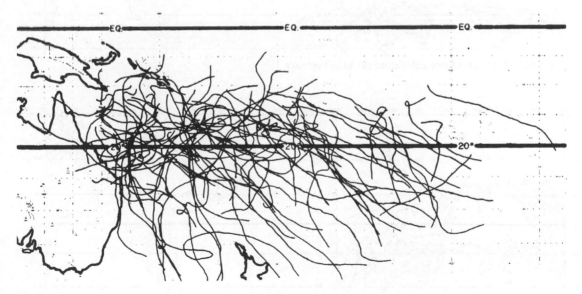

Figure 1.3(g) Typical tropical cyclone tracks in the south Pacific, for the 10-year period of 1968–77

Source: Gray 1975

14

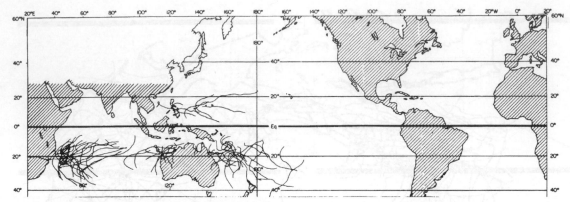

Figure 1.4(a) Ten years of tropical cyclone tracks in January

Source: Gray 1975

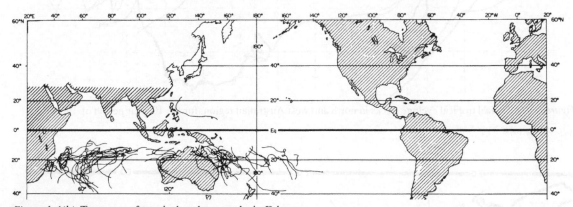

Figure 1.4(b) Ten years of tropical cyclone tracks in February

Source: Gray 1975

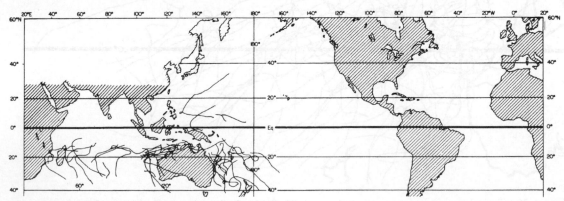

Figure 1.4(c) Ten years of tropical cyclone tracks in March

Source: Gray 1975

15

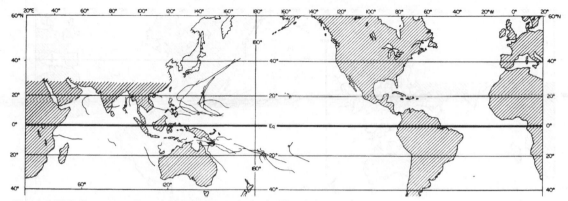

Figure 1.4(d) Ten years of tropical cyclone tracks in April

Source: Gray 1975

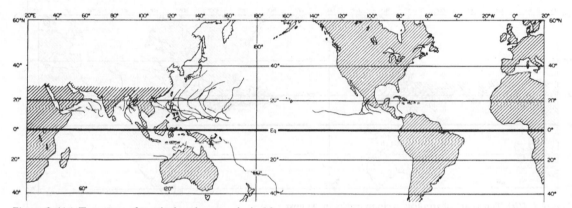

Figure 1.4(e) Ten years of tropical cyclone tracks in May

Source: Gray 1975

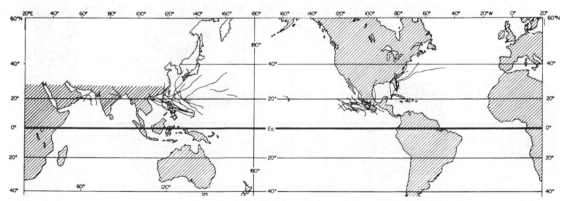

Figure 1.4(f) Ten years of tropical cyclone tracks in June

Source: Gray 1975

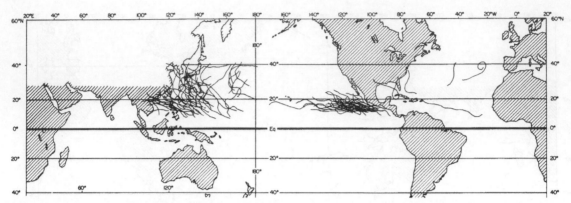

Figure 1.4(g) Ten years of tropical cyclone tracks in July

Source: Gray 1975

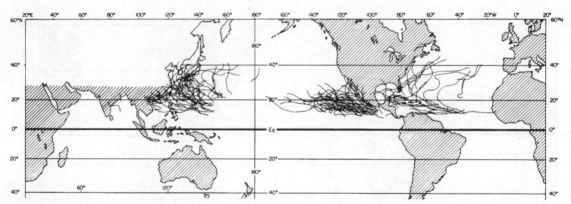

Figure 1.4(h) Ten years of tropical cyclone tracks in August

Source: Gray 1975

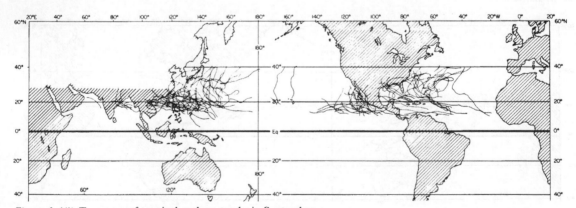

Figure 1.4(i) Ten years of tropical cyclone tracks in September

Source: Gray 1975

17

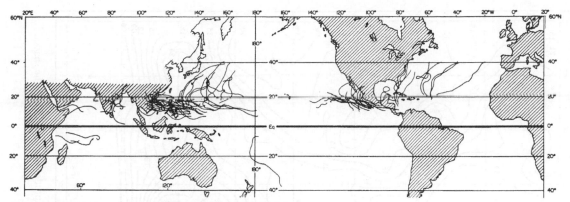

Figure 1.4(j) Ten years of tropical cyclone tracks in October

Source: Gray 1975

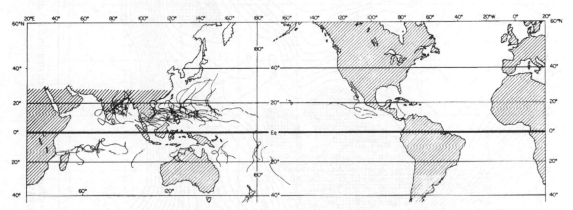

Figure 1.4(k) Ten years of tropical cyclone tracks in November

Source: Gray 1975

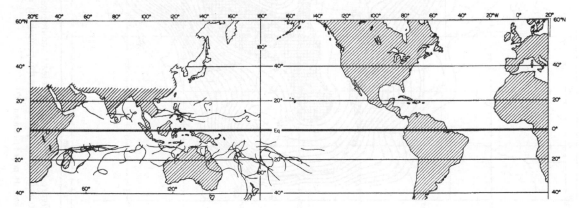

Figure 1.4(l) Ten years of tropical cyclone tracks in December

Source: Gray 1975

18

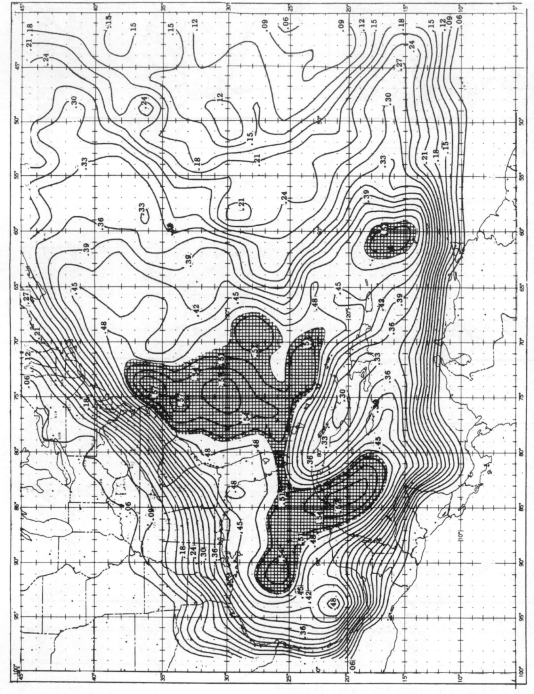

Figure 1.5 Probability of at least one tropical storm or hurricane per season entering unit 2.5 degree latitude–longitude boxes (the shaded areas indicate regions where this probability exceeds 0.5)

Source: Hope and Neumann 1971

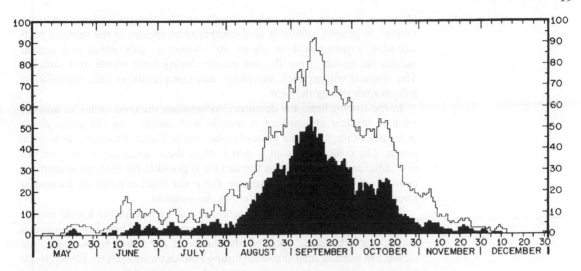

Figure 1.6 Number of tropical storms and hurricanes (open bar) and hurricanes (solid bar) observed on each day, 1 May to 30 December, 1886–1977

Source: Neumann *et al.* 1978

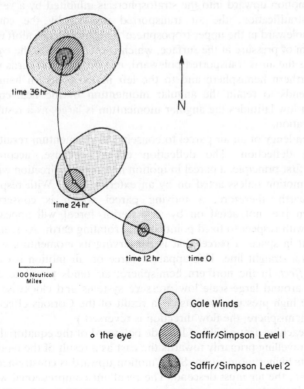

Figure 1.7 Use of the Saffir/Simpson scale to delineate storm intensity

However, because of the oblique angle of the sunlight, the temperatures remain, in general, relatively cool compared to regions in the summer mid-latitudes. Equatorward of about 30°, however, substantial and similar radiational heating from the sun occurs during both winter and summer. The tropical troposphere, therefore, has comparatively little variation in temperature during the year.

In the troposphere, the demarcation between the cold, polar air and the warmer tropical atmosphere is usually well defined by the *polar front*: poleward of the front, the air is of polar origin; equatorward it is of tropical origin. The colder, polar air is denser than the tropical air. Over a 30 per cent difference in densities at the surface is possible for extreme wintertime contrasts. During the winter season, the polar front is generally located at lower latitudes and is stronger than in the summer.

The region of greatest solar heating at the surface in the humid tropics results in areas of deep cumulonimbus convection. These cumulonimbus clouds occur because, upon condensation in the lower troposphere, the clouds are warmer than the surrounding ambient atmosphere. These clouds transport water substance, sensible heat and the earth's rotational momentum to the upper portion of the troposphere. The tropopause in these latitudes is around 17–18 km as a result of the vigorous mixing of the atmosphere by the convection.

Since motion upward into the stratosphere is inhibited by a very stable thermal stratification, the air transported upward by the convection diverges poleward in the upper troposphere. This divergence aloft results in a minimum of pressure at the surface, which is referred to as the *equatorial trough*. As the air is transported poleward, it is deflected towards the right in the northern hemisphere and to the left in the southern hemisphere, since it tends to retain the angular momentum of the near-equatorial region. At low latitudes the angular momentum is large, as a result of the earth's rotation.

(The tendency of an air parcel to conserve its momentum results in this horizontal deflection. The deflection occurs because, according to Newton's first principle, a parcel in motion in a certain direction will retain the same motion unless acted on by an exterior force. With respect to a rotating earth, therefore, a moving parcel which is conserving its momentum [i.e. not acted on by an exterior force] will appear to be deflected with respect to fixed points on the rotating earth. As seen from a fixed point in space, a parcel that is conserving its momentum would be moving in a straight line. This apparent force on air motion is called the *Coriolis effect*. In the northern hemisphere, air tends to rotate counter-clockwise around large-scale low pressure systems and clockwise around large-scale high pressure systems, as a result of the Coriolis effect. In the southern hemisphere, the flow direction is reversed.)

Upon reaching around 30° of latitude poleward of the equatorial trough, the air is travelling primarily towards the east as a result of the tendency to conserve its higher momentum. Since motion upward is constrained by the stratosphere, the air must descend. The resultant compressional warming, as the air descends, creates vast regions of strong, thermodynamic stability within the troposphere. The sparse precipitation in these regions, a result of thermodynamic stabilization and subsidence, is associated with the great arid regions of the world such as the Sahara, Atacama, Kalahari, and Sonoran deserts. The accumulation of air as a result of the convergence in the upper troposphere causes deep high pressure systems, referred to as *subtropical ridges*, to occur in these regions. Locally, these ridges or high pressure systems are given names such as the Bermuda High, the Azores High, and the North Pacific High.

Upon reaching the lower troposphere, the presence of the earth's surface requires that the air diverge, with some air moving poleward, the remainder equatorward. In either direction, the air is deflected to the right in the northern hemisphere and to the left in the southern hemisphere by the Coriolis effect. In the flow moving equatorward, this deflection results in north-east winds north of the equator and south-east winds south of that latitude. These low-level winds are called the *trade winds* since, in the seventeenth and eighteenth centuries, sailing vessels used them to travel to the Americas. The low-level convergence region of the north-east and south-east trade winds from the two hemispheres is called the *intertropical convergence zone* (ITCZ). The ITCZ corresponds to the equatorial trough and is the mechanism which helps to generate the deep thunderstorms in this surface pressure minimum.

The circulation of ascent in the equatorial trough, poleward movement in the upper troposphere, descent in the subtropical ridges, and equatorward movement in the trade winds is a *direct heat engine*, called the *Hadley cell*. It is a persistent circulation feature whereby heating from the latitudes of greatest insolation are transported to the latitudes of the subtropical ridges. The geographic location of the Hadley circulation moves north and south with the seasons, with the equatorial trough lagging the latitude of greatest surface solar heating by about two months. This lag results because of the thermal inertia of the earth's surface, in which the average highest surface temperatures are achieved after the time of greatest insolation, since time is required to heat the ocean surface waters and the soil. Off the east coast of the United States, for example, maximum ocean surface water temperatures occur in August and September, not in June during the solstice.

Poleward of the subtropical ridges in the lower troposphere, as a result of the tendency for air motions to conserve absolute angular momentum, south-westerlies in the northern hemisphere and north-westerlies in the southern hemisphere tend to occur.

Since warm air is being moved poleward at low levels, however, the wind flow is no longer associated with a direct heat engine. The heat which originated in the equatorial trough is consequentially transported nearer to the pole by large, nearly horizontal low pressure eddies which are called *extratropical cyclones*. These develop on the polar front when a sufficiently large horizontal gradient of temperature in the lower troposphere develops across the front. The intensity of this temperature gradient is referred to as the *baroclinicity* of the front.

Extratropical cyclones are found to have three stages of development:

the *developing stage* in which an undulating surface low pressure area develops along the polar front; the *mature stage* in which sinking cold air sweeps equatorward, west of the surface low, and ascending warm air moves poleward, east of the cyclone; and the *occluded stage* in which the warm air has been entrained within and moved above the air of polar origin and cut off from the source region of the tropical air. Cyclones which evolve no further than the developing stage are referred to as *wave cyclones*, while extratropical lows that reach the mature and occluded stages are *baroclinically unstable waves*. Extratropical storm development is referred to as *cyclogenesis*. The occurrence of surface pressure falls of greater than about 24 mb per day, which occasionally accompany rapid extratropical cyclone development, is referred to as *explosive cyclogenesis* and is often associated with major winter storms. Theoretical analysis has shown that the occurrence of baroclinically unstable waves is directly proportional to the magnitude of the temperature gradient, with maximum growth for wavelengths of 3,000–5,000 km.

Cold fronts occur at the leading edge of the polar air moving equatorward, while *warm fronts* are defined at the equatorward surface position of the polar air as it retreats poleward, east of the extratropical cyclone. The air moving equatorward behind the cold front occurs in pools of cold, dense, polar and arctic surface high pressure systems. Arctic highs are defined to distinguish air originating even deeper within the high latitudes than that of polar highs. When the polar air is neither retreating nor advancing, the polar front is called a *stationary front*. In the occluded stage, where the cold air west of the surface low pressure centre advances more rapidly eastward around the cyclonic circulation than the cold air east of the centre moves poleward, the warm, less dense tropical air is forced above the regions of cold air. The resultant frontal intersection is called an *occluded front*. Fronts of all types always travel in the direction towards which the colder air is moving.

Clouds, and often precipitation, occur poleward of the warm and stationary fronts whenever less dense tropical air moving poleward, north of subtropical ridges, reaches the latitude of the polar front and is forced upward over the colder air near the surface. Such fronts are defined as *active* fronts. Rain and snowfall from these form a major part of the precipitation that is received in mid-and high latitudes, particularly during the winter.

The position of the polar front slopes towards the colder air with height. This occurs because cold air, being more dense, tends to undercut the warmer air of tropical origin. Since the cold air is more dense, pressure decreases more rapidly with height on the polar side of the polar front than on the warmer side. In the middle and upper troposphere, the resultant large horizontal pressure gradient between the polar and tropical air creates strong westerly winds as air circulates around the region of low pressure in the higher latitudes at these heights. The centre of this low pressure region is called the *circumpolar vortex*. The region of the strongest winds, which occurs at the frontal juncture of the tropical and polar air masses, is called the *jet stream*. Since the temperature contrast between the tropics and the high latitudes is greatest in the winter, the jet stream is stronger during that season. In addition, since the mid-latitudes also become colder during the winter, while tropical temperatures are relatively unchanged, the westerly jet stream tends to move equatorward during the colder season.

The jet stream reaches its greatest velocities (up to 320 km/h) at the tropopause. Above that level, tropopauses in the polar region, which are lower in the tropics, result in a reversal in the stratosphere of the horizontal temperature gradient that is found in the troposphere, with relatively warmer temperatures at high latitudes. This causes a weakening of the westerlies with height. At intervals of 20–40 months, with a mean of 26 months, a reversal of wind direction occurs at low latitudes in the strato-sphere, such that an easterly flow develops. This feature is called the *quasi-biennial oscillation.*

An idealization of the general wind circulation within the troposphere, as discussed above, is illustrated in Figure 1.8 for the situation without continents.

Preferred geographic locations exist for the development, movement and decay of extratropical cyclones, and for the presence of centres of the subtropical ridge. In the winter, in mid-and high latitudes, continents tend to become lower tropospheric high pressure reservoirs of cold air as heat is radiated out to space during the long nights. In contrast, oceans lose heat less rapidly as a result of the large thermal inertia of water, its ability to overturn as the surface cools and to become negatively buoyant, and the existence of ocean currents such as the Gulf Stream and Kuroshio Currents, which transport heat from lower latitudes poleward. During the winter, the lower troposphere over the warmer oceanic areas therefore tends to become a region of relative low pressure. As a result of this juxta-position of cold and warm air, the east sides of continents and the western fringes of oceans in mid- and high latitudes are preferred locations for extratropical storm development. Over the Asian continent, the cold high

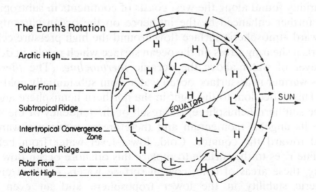

Figure 1.8 Schematic of the general circulation of the earth in the northern hemisphere winter. The sun would be located well to the right of the figure. There is average subsidence in the subtropical ridge and arctic high, and average ascent in the intertropical convergence zone and the polar front region.

pressure system is sufficiently permanent that a persistent offshore flow called the *winter monsoon* occurs.

An inverse type of flow develops in the summer as the continents heat more than adjacent oceanic areas. Continental areas tend to become regions of relative low pressure, while high pressure in the lower troposphere becomes more prevalent offshore. Persistent lower tropospheric onshore flow, which develops over large land masses as a result of the heating, is referred to as the *summer monsoon*. The leading edge of this monsoon is associated with a trough of low pressure called the *monsoon trough*. Tropical moisture brought onshore by the monsoon often results in copious rainfall. Cherrapunji in India, for instance, recorded over 9 m of rain in one month (July 1861) owing to the Indian summer monsoon.

The subtropical ridge is segmented into surface high pressure cells as a result of the continental effect. In the subtropics, large land masses tend to be relative centres of low pressure as a result of the strong solar heating. Persistent high pressure cells, therefore, such as the Bermuda and Azores Highs, occur over the oceans. The oval shape of these high pressure cells causes different thermal structures in the lower troposphere on their eastern and western sides. On the east, subsidence from the Hadley circulation is enhanced as a result of the tendency for air moving equatorward to preserve its angular momentum on the rotating earth. On the western side of the high pressure cell, air moving poleward must ascend in order to preserve its momentum. As a result of the enhanced descent in the eastern oceans, land masses adjacent to these areas tend to be deserts, such as found in north-west and south-west Africa, and along western coastal Mexico. In contrast, despite being under the descending branch of the Hadley cell, western continental fringes of the subtropical oceans are more likely to have precipitation, since the stablization effect of the subsidence portion of the Hadley cell is minimized by the upward vertical velocity that is associated with the western side of circular subtropical high pressure cells.

The aridity found along the west coasts of continents in subtropical latitudes is further enhanced by the influence on the ocean currents of the equatorward atmospheric surface flow around the high pressure cells. This flow exerts a shearing stress on the ocean surface which results in deflection of the layer of water above the oceanic *thermocline*. (The *thermocline* separates warmer near-surface ocean water from substantially colder water below.) This deflection, to the right in the northern hemisphere and to the left in the southern hemisphere, is a result of the tendency of the water to conserve its angular momentum and therefore to move westward when displaced toward the equator. Cold, lower-level water from below the thermocline rises to the surface to replace this offshore ocean flow. Called *upwelling*, these areas of cold, coastal surface waters result in enhanced atmospheric stability in the lower troposphere and an even greater reduction in the likelihood of precipitation, although fogs and low stratus clouds are common. Upwelling regions are also associated with enriched sea life because the cold, oxygen- and nutrient-rich bottom ocean waters are transported up near to the surface.

North-south oriented mountain barriers and large massifs, such as the Rocky Mountains, Scandinavian Mountains, and Tibetan Plateau, also influence the atmospheric flow. By imposing a barrier to the general westerly flow in mid-latitudes, they tend to block the air and to transport it poleward west of the terrain and equatorward east of the obstacle. Air that is forced up the barrier is often sufficiently moist to produce considerable precipitation on windward mountain slopes, while subsidence on the lee slopes produces more arid conditions. The elevated terrain affects the atmosphere as if it were an anticyclone, with the result that warm air is transported further towards the pole west of the terrain. It is also difficult for cold air in the interior to move westward of the terrain, therefore relatively mild weather for the latitude exists, for example, along the west coast of North America. In contrast, east–west mountain barriers, such as the Alps, offer little impediment to the general westerly flow, resulting in maritime conditions extending far inland.

A major focus of weather forecasting in the mid- and high latitudes is to predict the movement and development of extratropical cyclones, polar and arctic highs, and the location and intensity of subtropical ridges. Spring and autumn frosts, for example, are associated with the movement equatorward of polar highs behind a cold front, while droughts and heat waves in the summer are associated with unusually strong subtropical ridges. In the tropics, the location, intensity and structure of the inter-tropical convergence zone (ITCZ), the subtropical ridge, and monsoon troughs are critical components in weather forecasting. The disastrous drought in the Sahel in the 1970s, for example, was associated with a less than normal movement of the ITCZ to the north during the northern hemisphere summer, as well as a below average number of tropical cyclones which developed from weather systems moving westward off the coast of Africa.

Tropical cyclones appear to develop as a mechanism to transport heat and moisture poleward when the Hadley cell is unable to accomplish this exchange rapidly enough. They almost always develop on the equatorward or west side of the subtropical ridges over warm ocean waters. Frequently, as seen in Figures 1.3(a)–(g) and 1.4(a)–(l) and Appendix A, tropical cyclones move poleward around the west side of the subtropical high pressure regions. As these cyclones move poleward, they are absorbed into extratropical cyclones along the polar front, thereby completing the exchange of heat and moisture between the tropics and polar regions.

TROPICAL CYCLONE DEVELOPMENT CRITERIA

Observations have shown that necessary conditions for the development of tropical storms and hurricanes generally are:

(a) An underlying ocean with surface temperatures above 26°C (79°F).
(b) Small wind speed and direction changes with height between the lower and upper troposphere (a vertical wind shear of the horizontal wind between these two levels of less than 15 kn.)

(c) The presence of a pre-existing region of lower tropospheric horizontal wind convergence.

(d) A distribution of temperature with height which will overturn when saturated, resulting in cumulonimbus clouds.

(e) A location poleward of about 4°–5° of latitude.

From the discussion of the general circulation of the earth, presented in the previous section, locations around the world in which these conditions are met during at least some portions of the year are:

(a) Significantly equatorward of the polar front.

(b) Oceanic regions with sufficiently warm ocean temperatures, at least during the summer season.

Figures 1.9(a)–(d) illustrate regions of ocean surface temperatures greater than 26°C, while Figures 1.10(a)–(d) document the seasonal average wind flow for one year at 850 mb and 150 mb. The core of strong westerlies at 850 mb (850 mb generally occurs around 1.5 km above sea level with lower heights towards the poles) approximately corresponds to the location of the polar front. A wind shear of greater than about 15 kn between 850 mb and 150 mb does not favour the development of tropical cyclones.

Even a qualitative examination of Figures 1.8 and 1.10(a)–(d), with respect to the development criteria listed on pp. 25–6, and a comparison with the genesis regions of tropical cyclones, which are plotted in Figures 1.1 (a)–(d) and 1.2(a), (b), demonstrates clearly the validity of these criteria.

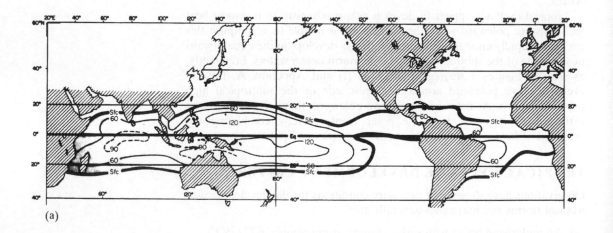

(a)

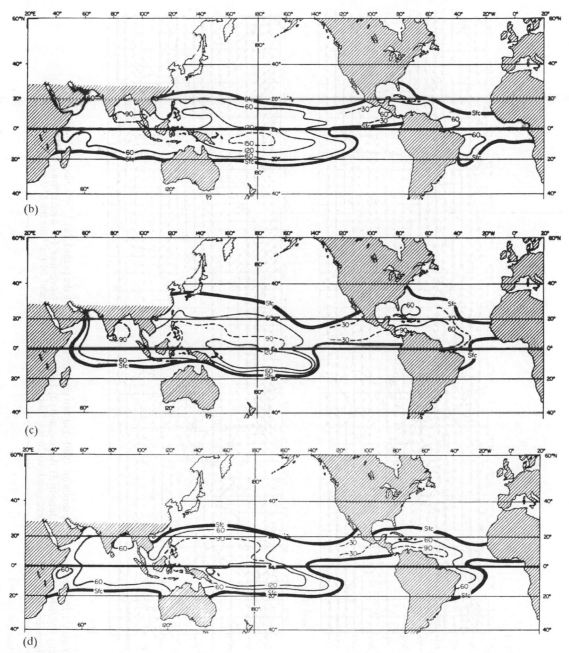

Figure 1.9 Seasonal climatology on the depth of the 26°C isotherm: (a) January–March, (b) April–June, (c) July–September, (d) October–December

Note: Sfc indicates that the isotherm is at the surface

Source: Gray 1981

28

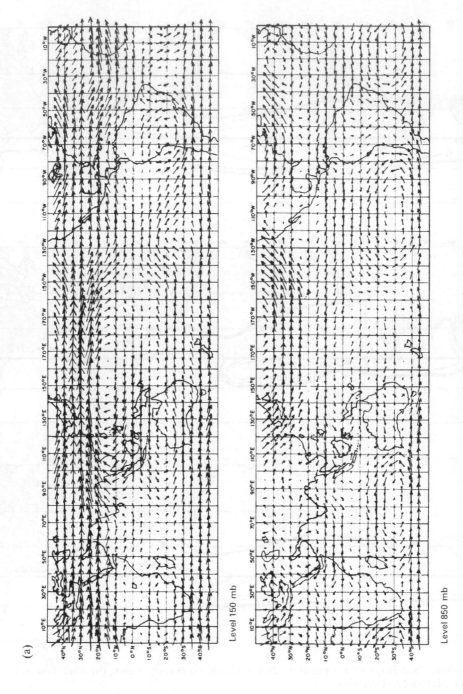

(a)

Level 150 mb

Level 850 mb

Figure 1.10 Seasonal average winds for 1980–1 at two pressure levels during the periods (a) 1 December to 28 February, (b) 1 March to 31 May, (c) 1 June to 31 August, and (d) 1 September to 30 November. For wind arrows, the distance between two vertical lines corresponds to 28 ms⁻¹ for the 150 mb Figures and 14 ms⁻¹ for the 850 mb Figures.

Source: White 1982

(b)

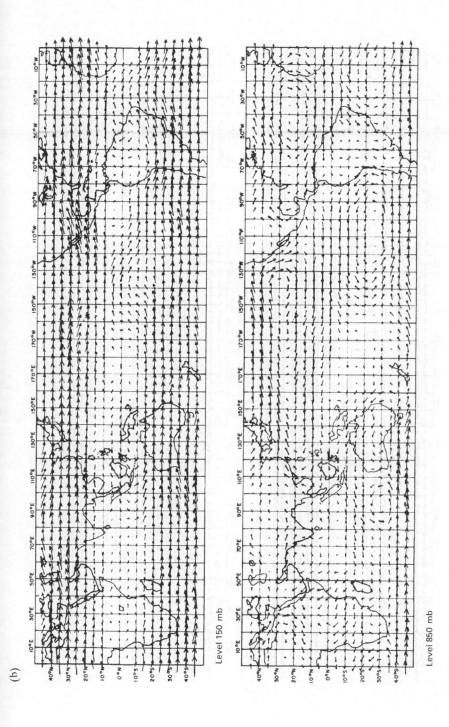

Level 150 mb

Level 850 mb

(c)

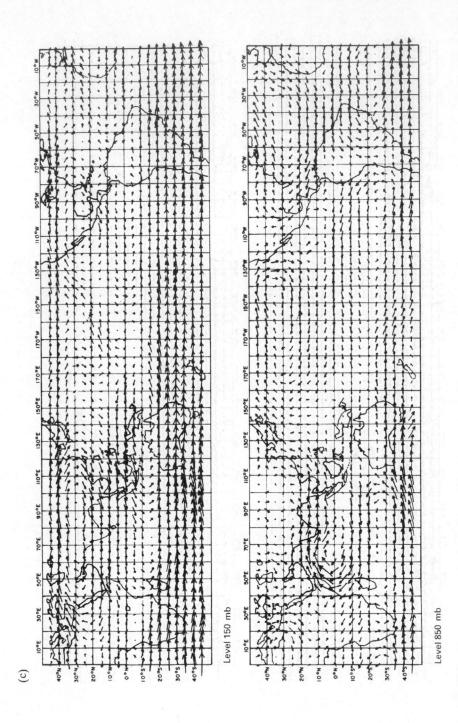

Level 150 mb

Level 850 mb

31

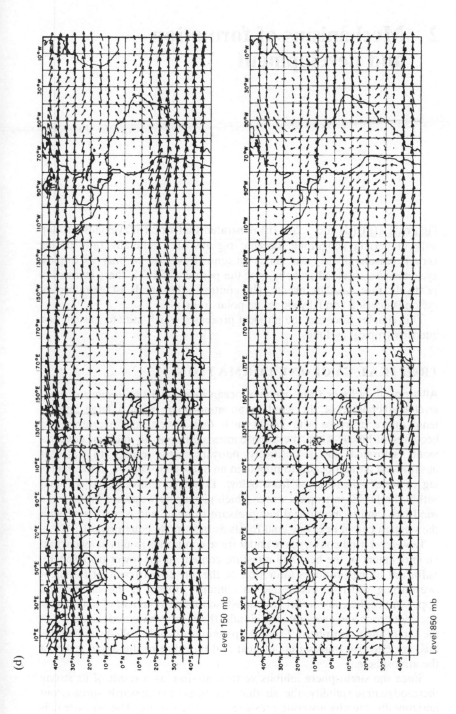

(d)

Level 150 mb

Level 850 mb

2 Mechanisms of formation and development

Figures 1.1(a)–(d) and 1.2(a),(b) illustrated that tropical cyclones develop only in certain geographic oceanic regions and generally only during certain seasons of the year. Since an annual average of only 99 tropical cyclones was observed globally over the period 1952–71, these storms are relatively rare weather events. (In contrast, thousands of extratropical cyclones develop annually along the polar front.)

This chapter discusses the physical processes which initiate, sustain and intensify tropical cyclones.

TROPICAL CYCLONE FORMATION

All tropical cyclones form over the ocean in pre-existing regions of low-level convergence of the wind in an atmosphere with a large enough temperature decrease with height such that it vertically overturns and becomes turbulent when the air becomes saturated. In such an environment, saturated air is positively buoyant with respect to adjacent unsaturated air and begins to bubble in an analogous fashion to the bubbling of water when heated from below. Low-level convergence is found within the ITCZ, in tropical waves which propagate westward in the trade wind belt, and in monsoonal low pressure troughs, and is associated with the western end of decaying cold fronts as they move equatorward.

Low-level wind convergence under these conditions results in ascent and an enrichment of the heat and moisture content of the lower troposphere. Sufficient lifting produces saturation of the air, and a resultant turbulent overturning which is manifested as cumulus clouds. If the middle and upper troposphere are sufficiently moist, so that entrainment of ambient air does not erode and evaporate the clouds, these cumulus clouds can grow to the top of the troposphere. These deep cumulus clouds, called cumulonimbus, are effective transporters of heat and moisture to the upper levels of the atmosphere.

Since the stratosphere inhibits vertical motion, as a result of its strong thermodynamic stability, the air that is transported upwards spreads out horizontally, thereby lowering pressures near the surface. The pressure falls near the surface result in an enhancement of the low-level wind convergence, thereby providing an impetus for additional cumulonimbus development.

This process of low-level wind convergence, ascent and heating in the middle and upper troposphere, plus upper-level wind divergence, is a form of thermal heat engine. The heating results from the release of *latent heat* as cumulonimbus develop and water vapour is converted to liquid water and ice. (Latent heat is the heat energy that is required for a change of phase of a chemical. For water, heat is released when water vapour condenses into liquid water. In contrast, heat is absorbed when liquid water evaporates into water vapour. Since the heat is realized only during a change of phase, it is referred to as latent heat.) Pressure will continue to fall at the surface in the centre as long as the divergence aloft is greater than the convergence at low levels. This process is illustrated schematically in Figures 2.1(a) and (b). Generally, surface pressure does not fall very far because the air which diverges aloft sinks at the periphery of the cumulonimbus cloud system and is recycled into the storm. This results in halting the pressure drop if the air is recycled and none of it is transported far from the region of deep convection. In addition, the subsiding air warms and dries the region surrounding the cloud system, providing an atmosphere which is less conducive to vigorous, deep cumulus convection.

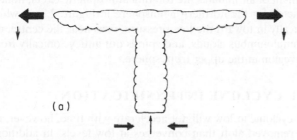

(a)

Figure 2.1(a) Development of deep cumulonimbus convection from low-level horizontal wind convergence

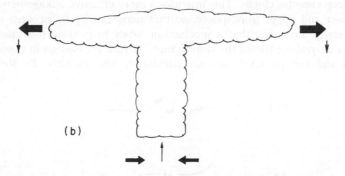

(b)

Figure 2.1(b) The cumulonimbus region will tend to become a broader area of deep convection if enhanced low-level wind convergence is caused by a pressure fall below the core of the cloud system because more mass is removed aloft than is replaced at low levels

At and near the equator, this representation of areas of deep cumulonimbus convection which occur without substantial pressure falls and without strong, sustained, low-level convergent winds is realistic, and occurs frequently within the ITCZ when it is located nearer than about 4° or 5° from the equator.

Poleward of that latitude, however, as air moves toward the low pressure near the surface and out from the high pressure region aloft, the tendency of an air parcel to conserve its momentum results in a horizontal deflection of the parcel, to the right in the northern hemisphere and to the left in the southern hemisphere, as discussed on p. 20. This Coriolis effect becomes larger towards the poles, since the reduction in distance between the earth's troposphere and the earth's axis of rotation becomes less, resulting in greater apparent deviations from horizontal, straightline motion as viewed with respect to a co-ordinate system defined on the rotating earth.

The result of the Coriolis effect poleward of about 4°–5° of latitude is that air which moves towards low pressure begins to turn cyclonically (which is counter-clockwise in the northern hemisphere, clockwise in the southern hemisphere), while the divergent outflow air aloft moves anticyclonically.

This movement of air initiates the rotating atmospheric circulation, schematically depicted for the northern hemisphere in Figure 2.2, in which air spirals cyclonically in toward the low pressure, rises near the centre, resulting in deep cumulonimbus clouds, and spirals out anticyclonically from the high pressure region in the upper troposphere.

TROPICAL CYCLONE INTENSIFICATION

The low-level cyclonic inflow will not accelerate with time, however, unless more mass is removed aloft than converges at low levels. In addition, the moisture enrichment of the middle and upper troposphere should continue, so that subsequent cumulonimbus can more easily grow to near the tropopause level, resulting in a greater percentage coverage of the disturbed area with deep cumulus clouds. This provides a more effective linkage between the lower and upper troposphere, so that mass can be transported aloft more easily. Providing that a mechanism exists to exhaust this mass to regions far removed from the disturbance, surface pressures will continue to fall and the low-level cyclonic circulation will intensify. Dr Robert

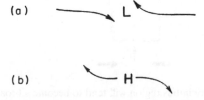

Figure 2.2 Schematic of flow (a) towards a low pressure and (b) outwards from a high pressure centre which is deflected to the right by the Coriolis effect

Merrill, in his Ph.D. dissertation at Colorado State University (Merrill 1985), concluded that divergent upper-level winds, at and beyond the periphery of the storm outflow region, are required to remove the mass from the storm environment, and thus to prevent a recirculation of the air back into the storm at low levels. Thus the large-scale flow at 200 mb (around 12 km) is critical as to whether or not a tropical cyclone continues to intensify.

As the low-level flow becomes stronger and a tropical cyclone develops, it becomes increasingly more difficult for air to reach all of the way into the centre. This limitation occurs because air parcels achieve larger rotation rates as they spiral into the centre, resulting in a tendency for the parcel to be spun out of the inward spiral. Referred to as the *centrifugal force* (or, alternatively, the *centripetal acceleration*), this is the effect that makes it increasingly more difficult to rotate a ball on a rope around oneself, while shortening the rope.

The spiralling in of airflow at low levels results in the formation of cumulonimbus spiral bands, which are concentrated in localized regions of the curved inflow. The inflow becomes enhanced in several spiral bands because deep cumulus clouds, as they are generated, enhance the low-level flow in their vicinity, resulting in a more favourable environment of moisture and temperature for deep convection than that found in the inflow region without clouds. This positive feedback between enhanced local convergence and cumulonimbus bands in the inflow perpetuates the spiral bands. The spiral bands tend to rotate slowly cyclonically, at less than the speed of the wind, around the centre of the tropical cyclone. These spiral bands are also called *feeder bands* because of their contribution of heat and moisture to the centre of the cyclone. Figure 2.3 illustrates from satellite imagery the form of these spiral bands as seen from space.

When the inward flow accelerates, such that minimal hurricane force winds are achieved, air parcels can no longer reach the centre of the low pressure, but blow tangentially to lines of constant pressure, some distance from the centre. This region, called the *eye wall*, corresponds to the maximum inward penetration of the inward spiralling air and is the region of strongest winds. Inside this ring of maximum winds (often referred to as the *radius of maximum winds*), a more or less becalmed area occurs, called the *eye*. A major distinction between a tropical storm and a hurricane is the occurrence of an eye when the sustained winds near the centre reach about 64 kn or greater. This development of an eye is schematically illustrated in Figure 2.4. Figure 2.5 illustrates the appearance of an eye as viewed from a geostationary satellite. Note the well-defined hole in the centre of the cloud mass which is associated with the typical cyclone. A computer simulation of flow into, upward, and out from the eye wall is shown in Figure 2.6. Figure 2.7 presents a view of the eye of a hurricane (Hurricane Donna, 1960) as seen from a radar. The radar senses rainfall, hence the rain-free area of subsidence in the eye is clearly evident in the photograph.

Inside the eye, the subsidence causes compressional warming of the atmosphere. Since warmer air can contain larger amounts of water vapour before condensation must occur, clouds tend to dissipate in the eye. In

Figure 2.3 A geostationary satellite infra-red image of Hurricane Gloria on 25 September 1985 at 17.00 GMT. The photograph is a representation of temperature, with the coldest clouds associated with the storm appearing in black.

addition, the warming owing to subsidence creates a thermodynamically more stable atmosphere which is less conducive to deep cumulus convection.

The subsidence can be explained as a response to lateral entrainment of air in the eye into the deep cumulonimbus in the eye wall and to low-level horizontal wind convergence from the eye into the updrafts below cloud base in the eye wall. The resultant loss of mass in the eye will cause the subsidence within the eye. The more rapid the loss of mass, the larger will be the magnitude of the subsidence. In intense and intensifying hurricanes, the eye region can be completely cloud-free. In very intense hurricanes such as Gilbert (1988), double, concentric eye walls can also form.

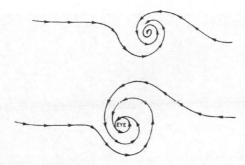

Figure 2.4 Schematic of the formation of an eye, as the winds and horizontal pressure gradient force become too strong to permit the air to spiral all the way into the centre of the tropical cyclone

A useful, simple formula exists to relate the maximum sustained wind in an Atlantic hurricane to the difference in sea level pressure between the eye, P_{EYE}, and at the surface at the periphery of the storm, P. If P_{EYE} and P are given in millibars, then in metres per second

$$V_{\text{maximum sustained 1-minute average wind}} = 6.3 \, (P-P_{EYE})^{1/2}$$

While variations from this formula occur, it has been found to be an extremely useful estimate. The application of this formula when $P = 1013$ mb is graphed in Figure 2.8. One of the lowest pressures ever observed in an Atlantic hurricane was in the Labor Day, Florida Keys Storm of 1935 in which 408 deaths occurred. Using the formula, the observed barometric pressure of 26.35 in. (892 mb) results in a maximum estimated one-minute averaged wind of 69 ms^{-1} (135 kn.), which is close to the observed peak winds of 150–200 statute m.p.h. (135 kn. corresponds to 155 statute m.p.h.)

The deepest pressure, and hence maximum wind, that is possible in a hurricane is limited by the ocean surface temperature. Since the hurricane is a direct heat engine, it makes intuitive sense that the wind circulation of such a storm will be stronger when the heating is greater. Figure 2.9(a) illustrates estimated minimum potential central pressure and maximum sustained winds as a function of sea surface temperature. The number of actual observed storms as a function of maximum measured wind speeds and minimum pressure, which was used to construct the maximum potential hurricane intensity, is also plotted in the Figure. From this Figure, level 5 Saffir/Simpson intensity hurricanes would not be expected if the ocean surface temperature were only 26°C (79°F). Figure 2.9(b) presents similar information relating maximum sustained wind to sea surface temperature over a wider temperature range.

In the eye, sinking air tends to occur, a consequence of divergent low-level flow and entrainment of air out from the centre and into the vigorous eye wall region of deep cumulonimbus convection, as discussed on p. 36. The result of this subsidence is a tendency for clear air within the eye. Individuals who experience the passage of the eye are often surprised by the appearance

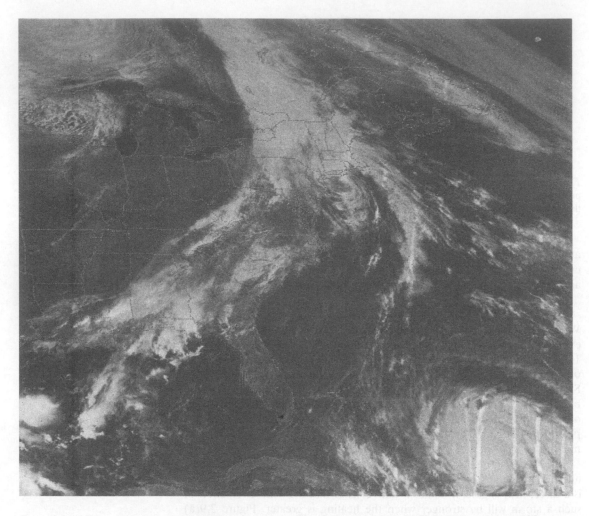

Figure 2.5 Visible satellite image of the hurricane eye in Hurricane Gloria at 1731 GMT on 24 September 1985

of blue, sunny skies (or stars at night) along with a dramatic cessation of the wind within the eye. Unfortunately, this interlude is usually followed by a rapid increase of the wind from the direction opposite to that observed before the passage of the eye.

Figure 2.10 illustrates the change of wind speed, temperature and 700 mb height from the periphery of a hurricane into its centre. This cross-section, obtained by averaging 16 aircraft penetrations into Hurricane Anita in the Gulf of Mexico on 2 September 1977 is typical of a strong, mature tropical cyclone. Note that the radius of maximum winds (of about 64 ms^{-1} [125 kn.]), which occurred about 20 km (11 nautical miles) out from the centre, corresponds to the outer limit of the warming of up to 5°C (9°F) in the eye. The magnitude of upward motion is also largest at the

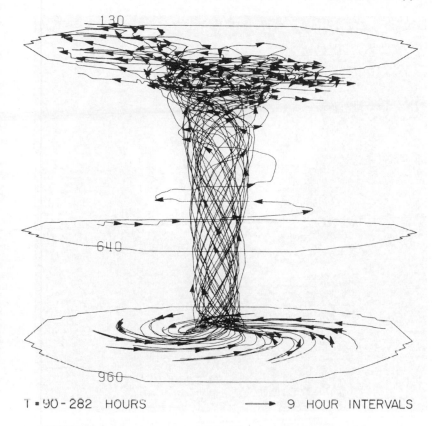

T ▪ 90 - 282 HOURS   ⟶ 9 HOUR INTERVALS

Figure 2.6 Particle trajectories calculated from a numerical model of an
asymmetrical hurricane

Source: Anthes and Trout 1971

radius of maximum winds with values up to 2.5 ms⁻¹ (5 kn.) The wind
perpendicular to the isobars (labelled as 'Rad wind' in Figure 2.10) is small
but must be inward on average in order to produce the concentrated region
of upward wind at the radius of maximum wind.

The view from within the eye can be spectacular in strong hurricanes.
The deep thunderstorm clouds of the eye wall have been characterized as
appearing like a gigantic rotating coliseum. Birds have been reported as
finding sanctuary within the storms, often being transported hundreds or
even thousands of kilometres from their native regions. Ships within the
eye often report numerous birds perching on their vessels in order to rest.

Robert H. Simpson graphically describes the appearance of the eye as
seen from an aircraft penetrating into Typhoon Marge:

Soon the edge of the rainless eye became visible on the (radar) screen.
The plane flew through bursts of torrential rain and several turbulent
bumps. Then suddenly we were in dazzling sunlight and bright blue sky.

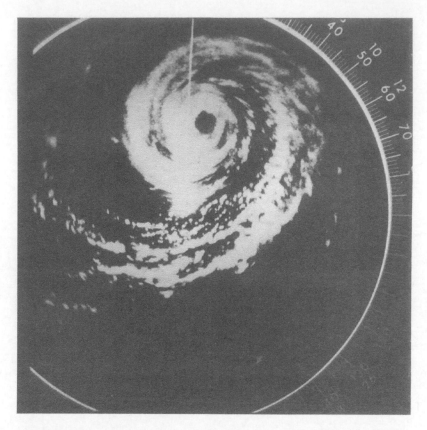

Figure 2.7 Radar picture of Hurricane Donna taken by the WSR-57 at Key West, Florida, 10 September 1960. The centre was about 65 miles (119 km) from the station.

Source: Dunn and Staff 1967

Around us was an awesome display. Marge's eye was a clear space 40 miles in diameter surrounded by a coliseum of clouds whose walls on one side rose vertically and on the other were banked like galleries in a great opera house. The upper rim, about 35,000 feet high, was rounded off smoothly against a background of blue sky. Below us was a floor of smooth clouds rising to a dome 8000 feet above sea level in the center. There were breaks in it which gave us glimpses of the surface of the ocean. In the vortex around the eye the sea was a scene of unimaginably violent, churning water.

(Simpson 1954)

Hurricanes are observed only over water or for short distances inland after they make landfall, before they weaken to below hurricane strength. There are four major reasons for the inability of these storms to maintain their intensity near the surface after landfall.

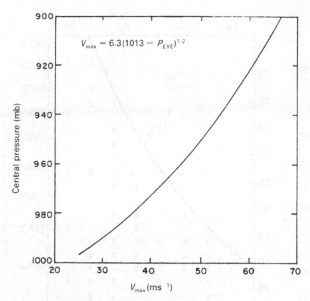

Figure 2.8 Approximate value of maximum sustained one-minute averaged wind speed in Atlantic hurricanes as a function of central pressure

Source: Adapted from Simpson and Riehl 1981

First and most important, as already discussed, a hurricane is a direct thermal heat engine. This requires that the warmest temperatures which are associated with the storm be in its centre. However, as air spirals into a hurricane, it expands as a result of the lower pressures closer to the eye. Unless heat is added, this expansion results in cooling. (The same process occurs when air is let out of a tyre. The expanding air at the nozzle from the pressurized tyre is substantially colder than the surrounding air.) The cooling works against maintaining the heat engine. For example, air which originates at a pressure of 1000 mb and 27°C (80°F) in the region surrounding the storm would cool to 18°C (64°F) at a pressure near the centre of a storm of 900 mb, unless heat were added.

Over land there is no heat source to counteract this cooling. The direct result is that deep cumulonimbus convection over land becomes inhibited as negatively buoyant cool air is advected into the eye wall region. The low central pressure of the tropical cyclone correspondingly rises as the coupling between the lower and upper troposphere is reduced and, as a result, the divergent winds in the upper levels of the tropical cyclone are diminished in strength. Thus, the eye wall tends to be destroyed as the hurricane weakens to tropical storm strength.

Over warm oceanic regions, in contrast, the water serves as a source of heat. When an air parcel starts to cool as a result of expansion, it tends to overturn because of the positive buoyancy created as cooler air overrides the warmer sea water. Therefore, with an oceanic temperature of 27°C, for

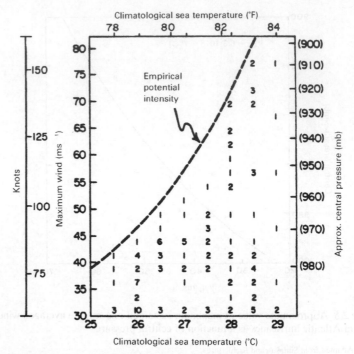

Figure 2.9(a) Estimate of maximum potential sustained wind speed and minimum central sea level pressure as a function of sea surface temperature

Source: Merrill 1985

example, an air parcel which originates at a pressure of 1000 mb and 27°C could retain that temperature when it reaches a pressure of 900 mb. The integrity of the heat engine and the favourable environment for the eye wall is not lost. It is important, however, that the ocean be as warm as the incoming air parcel; otherwise, the air will still cool by expansion over the cold water until the air and water reach the same temperature.

The warm ocean also serves as an essentially unlimited source of water to the deep cumulonimbus. This is the second reason that hurricanes occur only over oceanic regions. A surface of water evaporates at a rate which is directly related to its surface temperature. For the same amount of moisture just above the surface, and a surface pressure of 1000 mb, an ocean area with a temperature of 27°C, for example, will evaporate at a rate of about 64 per cent greater than when the surface is at 18°C.

Third, the development of sea spray in strong winds, and its subsequent evaporation, is an additional source of water vapour over the oceans. Over land, the availability of water is limited to the amount that may be extracted from the ground and plants through evaporation and evapo-transpiration, and to re-evaporation of rainfall on the ground. This evapor-ation over land also further aids to cool the lower levels of the atmosphere

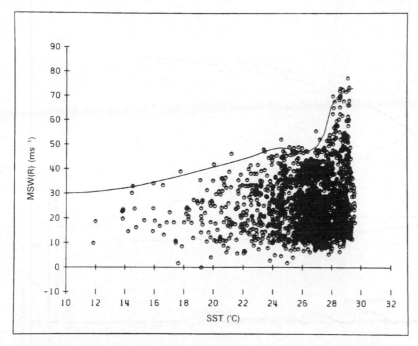

Figure 2.9(b) Scatter diagram of monthly mean sea surface temperature (SST) as related to maximum wind for a sample of north Atlantic tropical cyclones. The line is the 95 percentile and provides an empirical upper limit on intensity as a function of ocean temperature. The winds shown are maximum sustained (low-level) winds (MSW) relative (R) to storm motion.

Source: Merrill 1987

since heat is required (and lost to the air) in the conversion of liquid water to water vapour. Except for this meagre source of water, the water supply for cumulus activity over land is limited to existing water vapour in the atmosphere, either originally present or transported inland from adjacent marine areas.

A fourth substantial difference between the ocean and land areas is the generally larger aerodynamic roughness of the land. Trees, buildings, and even grasslands tend to be rougher surfaces, with the result that air is decelerated more over land than over the ocean. Even with large amplitude sea waves during windy conditions, the ocean remains relatively smooth aerodynamically, apparently a result of its ability to be moved by the wind. One major result of this difference in roughness is that, even if the wind at 100 m were the same, the greater retardation of the flow by the rougher surface over land would result in slower speeds at a height of a few metres above the ground.

The cooler surface over land, resulting from the expansion of the air and evaporation, magnifies this reduction in wind speed near the surface even

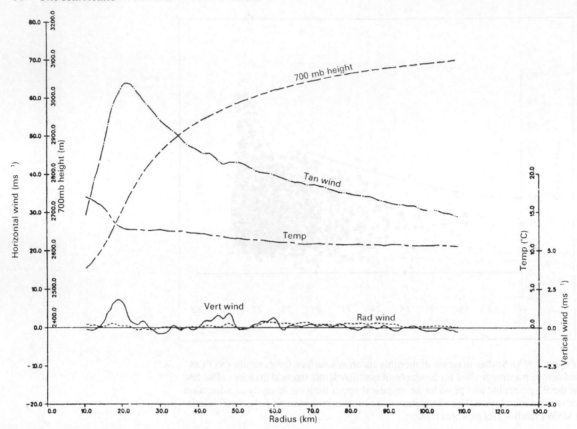

Figure 2.10 The azimuthal mean structure as approximated by the averaging of the sixteen profiles for Hurricane Anita on 2 September 1977

Source: Willoughby 1979

further, since the ability of the atmosphere to overturn and mix is inhibited by cooler air near the surface. Therefore, while the winds above about 100 m may accelerate, as those levels tend to become decoupled from the frictional retardation of the surface, the winds near the surface tend not to entrain the higher velocity air from aloft and, therefore, become quite weak. While this decoupling of near surface flow from the winds aloft does not directly reduce the overall intensity of a hurricane, its destructive potential near the ground is minimized. Figure 2.11 illustrates schematically the expected differences in wind profiles over the ocean and just inland that are associated with the landfall of a hurricane, as a result of a more stable temperature profile and rougher surface over land.

In summary, there are several major criteria for the development of tropical storms and hurricanes:

(1) The presence of a pre-existing synoptic-scale region of low-level convergence and low surface pressure.

(2) A warm, moist tropical atmosphere that is conducive to overturning when the air becomes saturated (i.e. favourable for cumulonimbus development).

(3) Oceanic surface temperatures greater than about 26°C (79°F) so that sufficient moisture and heat can be supplied to sustain the cumulonimbus.

(4) Weak vertical shear of the horizontal wind (less than about 15 kn. between the upper and lower troposphere within a radius of about 4° of latitude centred on the moving area of deep convection) such that maximum heating owing to the cumulonimbus remains over the region of lowest pressure.

(5) A distance sufficiently removed from the equator (generally by more than 4°–5° of latitude) such that air will tend to spiral inward cyclonically at low levels towards the lower pressure, and outward anticyclonically at upper levels away from high pressure.

(6) The development or superposition of a large-scale anticyclone in the upper troposphere over the surface low so as to evacuate mass far from the region of the cyclone, thereby permitting surface pressures to continue to fall.

Once a tropical cyclone reaches the intensity of a hurricane, it will not weaken unless

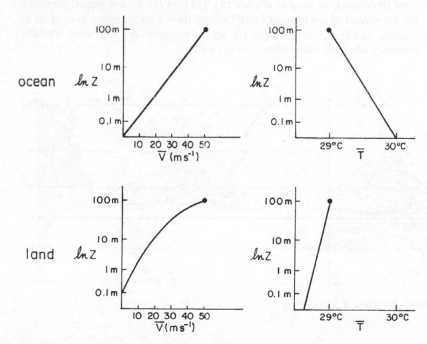

Figure 2.11 Schematic of differences in wind profiles and temperature over ocean and land owing to a relatively cool and rougher land surface

Note: z is the height above the surface

(a) its source of heat and moisture is reduced as a result of passage over land or relatively cold water,
(b) dry, cool air, which does not favour deep cumulonimbus convection, is transported into the system,
(c) the anticyclone aloft is replaced by a cyclonic circulation which adds mass to, rather than evacuates mass from, the hurricane heat engine. Since deep cumulonimbus convection tends to perpetuate an anticyclone aloft, larger scale atmospheric circulation changes are required to remove such an outflow region.

Note that the requirements for the continuance of a hurricane are less restrictive than those for its development. Therefore, while only occasionally are atmospheric conditions ripe for the genesis and development of a hurricane, once established it tends to be a persistent weather feature.

Figures 2.12(a) and (b) illustrate a procedure to quantify the necessary ingredients for tropical cyclone genesis and to relate it to the observed frequency of storms. The dynamic potential, as applied by Gray (1975), is determined from quantitative measures of criteria (1), (4) and (5), while the thermal potential is obtained from (2) and (3). Criterion (6) becomes particularly important in the transition of a tropical storm to a hurricane, and to further intensification.

In contrast to the genesis location, the ability of a hurricane to persist once developed, as long as criteria (a), (b) and (c) do not occur, accounts for the spread of storm tracks well beyond their source region as evident in Figures 1.3(a)–(g) and 1.4(a)–(l) and Appendix A. The next chapter discusses why and where these storms move.

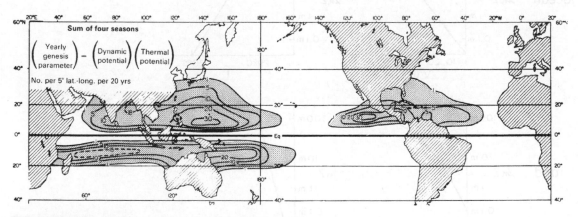

Figure 2.12(a) Favourable conditions for tropical cyclogenesis

Source: Gray 1975

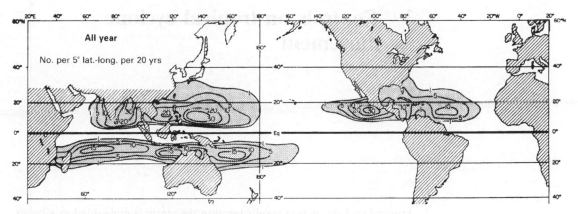

Figure 2.12(b) Observed locations of tropical cyclogenesis

Source: Gray 1975

3 Controls on tropical cyclone movement

Tropical cyclone motion results because the storm is embedded in a larger scale region of moving air, referred to as the *steering current*, which tends to move the low-level cyclonic, upper-level anticyclonic circulation and associated deep cumulonimbus convection in the direction of that flow. The cyclone itself, of course, is part of the large-scale flow, so that defining the appropriate steering current is difficult. The motion of the cyclone is also influenced by its own internal flow which, in general, is asymmetric with latitude variations across the storm (Holland 1983).

EXTERNAL FLOW

The definition of an appropriate steering current is not straightforward. G.J. Holland of the Bureau of Meteorology in Melbourne, Australia suggests using the winds averaged within a concentric band of 200–400 km from the storm centre. The averaging depth has been suggested to be between 500 mb and 700 mb (Chan and Gray 1982). Observations suggest that tropical cyclones typically move about 15° to the left and 20 per cent faster than a basic current defined in terms of a domain of 5°–7° of latitude from the centre.

Weather map analyses of the current and anticipated steering layer, therefore, provide considerable useful guidance for estimating the direction and speed of tropical cyclone motion.

If these currents were fixed in time, hurricane track forecasting would be comparatively simple. Unfortunately, this is not the case, as the orientation and strength of the steering current changes in response to the normal propagation and development of large-scale pressure ridges and troughs in the atmosphere.

The climatological flow fields at 850 mb at various times of a typical year were illustrated in Figures 1.10(a)–(d). Note the strong correlation between the flow directions in these Figures and the average tropical cyclone tracks as plotted in Figures 1.3(a)–(g) and 1.4(a)–(l) and Appendix A. Contrary to popular conception, most tropical cyclones have fairly regular, well-defined tracks because the climatological flow pattern occurs much of the time.

The difficulty in predicting storm track occurs either when the climatological pattern is replaced by a less common, large-scale flow or, more importantly, when rapid changes in time occur in the strength and orientation of the steering current.

For example, on 4 September 1965, Hurricane Betsy was moving northwest around the southern rim of the large Bermuda High subtropical ridge in the central Atlantic (see Figure 3.1). As the storm was moving northward off the east coast of the United States in a climatologically expected direction and speed, a readjustment occurred in the hemisphere flow patterns owing to a trough in the westerlies over the central United States. This change resulted in the propagation of the subtropical ridge towards the west until it was north of the storm system. As a result, Hurricane Betsy was blocked from continuing its climatologically expected northward movement and became stationary. The subtropical ridge centre continued to build westward so that, after about a day, the steering currents became northerly and the storm began to move south towards the northern Bahamas. With the re-establishment of the subtropical ridge centre to the west, the subsequent track of Betsy travelled around the new position of the Bermuda High, eventually slamming into New Orleans when it finally began once more moving northward around the western flank of the ridge. A major forecast problem associated with this storm was when it would begin its turn towards the west around the southern periphery of the high. An earlier turn would have brought Betsy onshore near Miami, with pos-

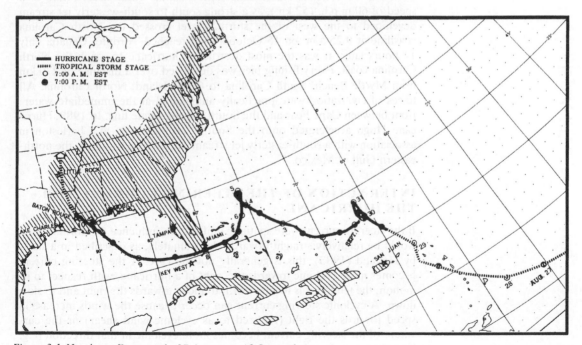

Figure 3.1 Hurricane Betsy track, 27 August to 12 September

Source: US Dept of Commerce 1965

sible major devastation to that urbanized area. A later turn would have permitted the storm to pass through the Florida Straits. As it happened, the storm crossed over the Florida Keys. Radar tracks of the storm are shown in Figures 3.2(a) and (b).

Therefore, while the news media attribute hurricanes 'with a life of their own', they are, of course, well-behaved natural phenomena and, to a large extent, their movement can be explained by the steering currents alone. The difficulty in forecasting their motion occurs when the steering currents are weak and ill-defined and/or when the future prediction of the steering currents is uncertain.

Tropical cyclones occasionally undergo rapid acceleration in forward motion. This happens when the storm becomes linked to a strong westerly jet stream which is associated with the polar front discussed on pp. 20–3. The tropical cyclone can become absorbed in a developing extratropical cyclone, infusing added moisture and resulting in a more intense extra-tropical storm than otherwise would occur.

In 1938, the development of a strong, south-westerly jet stream to the west of a hurricane resulted in the rapid acceleration of the hurricane to the north at a forward speed of up to 58 m.p.h. (50 kn.). The storm crossed Long Island, New York with little warning, resulting in 600 deaths in New England. Blue Hill, Massachusetts reported a five-minute average wind speed of 121 m.p.h. (105 kn.) with a gust to 183 m.p.h. (159 kn.).

In 1954, Hurricane Hazel underwent a similar rapid acceleration to a speed of 60 m.p.h. (52 kn.), as a strong south to south-westerly jet stream developed to the west of the storm. Hazel crossed the North Carolina coastline at 9.25 a.m. on 15 October, and reached Toronto, Canada only 14 hours later. It was the most destructive hurricane to reach the North Carolina coast. Every fishing pier was destroyed over a distance of 270 km from Myrtle Beach, South Carolina to Cedar Island, North Carolina. All traces of civilization were practically annihilated at the immediate water-front between Cape Fear and the South Carolina state line. In 1989, Hurri-cane Hugo accelerated onto the South Carolina coast at Charleston in association with a south-easterly jet stream caused by a trough in the north-eastern Gulf of Mexico.

INTERACTION OF THE STEERING CURRENT AND THE HURRICANE

If the steering current were spatially uniform, its influence on storm motion would be relatively straightforward. Unfortunately, as described in a study by Holland (1983), this is generally not the case. If the steering current becomes more cyclonic towards the right of a hurricane, with respect to its motion, the tendency is for the storm to move towards the right and to slow down. A cyclonic steering current can occur because the winds increase in speed towards the right (referred to as cyclonic wind shear) or the cur-vature of the steering current becomes more cyclonic in that direction (e.g. because a trough of low pressure is situated to the right of the storm). If, however, the steering current becomes more anticyclonic towards the right

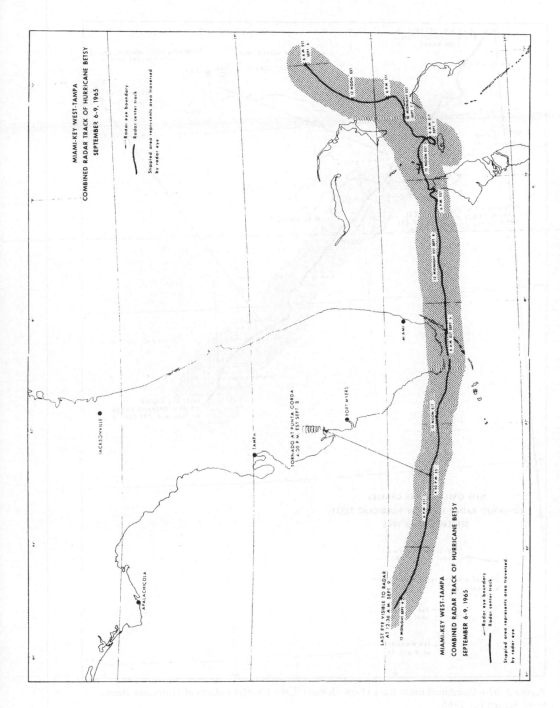

Figure 3.2(a) Combined radar track (Miami–Key West–Tampa radars) of Hurricane Betsy, 6–9 September 1965

Source: US Dept of Commerce 1965

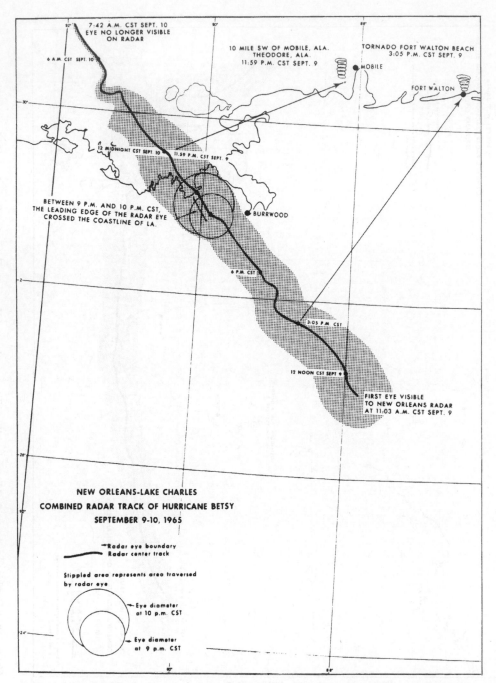

Figure 3.2(b) Combined radar track (New Orleans–Lake Charles radars) of Hurricane Betsy, 9–10 September 1965

Source: US Dept of Commerce 1965

of a moving hurricane, the tendency is for the storm to turn to the left. In addition, a downstream divergence of the steering current accelerates a storm, while a downstream convergence slows it down.

The spatial structure of the steering flow directly influences storm motion because the hurricane and steering current are not separate, distinct features but are intertwined with one another. The hurricane is not like a spinning cork flowing down a stream but is more analogous to an eddy within a stream. Just as with a hurricane, such a cyclonic eddy propagates towards a region in which the flow structure favours the generation of a cyclonic circulation.

A hurricane also moves because of the variation of the angular velocity of the earth across the storm. Since the influence of the earth's rotation on wind flow is greater at higher latitudes and since the Coriolis effect (see p. 20) acts to the right of the wind in the northern hemisphere and to the left in the southern hemisphere, then, in a symmetric wind speed field, flow moving equatorward on the west side of a storm causes a greater velocity towards the west than results from the air moving poleward on the east side of the hurricane. The net result is a tendency for a westerly drift with respect to the steering current.

The cumulonimbus associated with the hurricane also modifies the steering current. The mass which is evacuated from a hurricane in the upper tropospheric outflow descends at some distance from the storm. If this mass accumulation were symmetric around the storm, there would be no direct effect on the steering current. Observational analysis, however, shows that this outflow is often concentrated in what are called *outflow jets*, with the mass accumulation occurring in sub-regions of the environment surrounding the storm. If, for instance, this mass accumulation occurred in the front right quadrant with respect to the storm movement, the consequence would be a strengthening of the subtropical ridge from what would occur in the absence of the storm. The net result is a movement of the storm to the left of the track that it would have in the absence of this effect. (The neglect of this effect in the forecasting of the landfall location of Hurricane Gilbert in 1988 resulted in erroneous storm track predictions of a landfall on the Texas coast. Hurricane Gilbert made final landfall on the north-east coast of Mexico.) Correspondingly, an outflow jet which results in an accumulation of mass in the right rear quadrant would tend to accelerate the storm northward in the northern hemisphere.

The importance of the outflow mass accumulation on storm motion becomes more significant for larger and more intense storms for which the quantity of mass removal is greater. It also becomes a more important component in determining the track of a hurricane or tropical storm when the steering currents are weak and ill-defined. Otherwise, when the steering current is strong, the mass readjustment by the tropical cyclone itself is only of secondary importance in terms of determining storm motion.

Figure 3.3 presents a satellite picture to illustrate the asymmetric redistribution of mass by a hurricane outflow (Hurricane Gloria, 1985). The large avenue of bright clouds stretching to the south-east of the storm are cirrus clouds generated by cumulonimbus which were transported to that

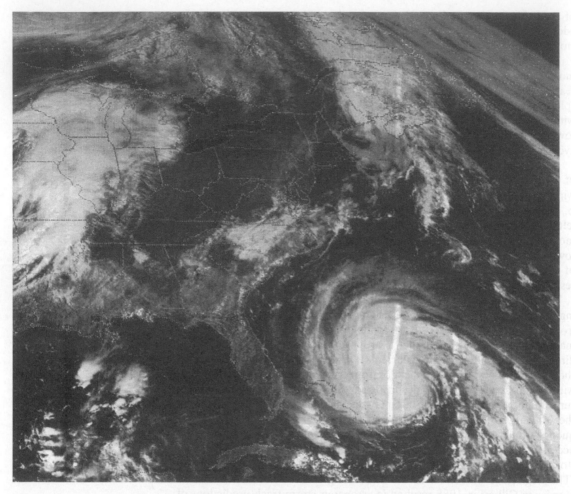

Figure 3.3 Satellite visible image of Hurricane Gloria at 1731 GMT on 25 September 1985, showing a cirrus outflow jet moving south-east from the storm

quadrant of the system by the outflow jet. These clouds will continuously dissipate on their downward edge as the air sinks in this region and helps to strengthen the subtropical ridge in that area.

Terrain also influences the movement of a hurricane. Figure 3.4 illustrates a numerical model simulation of the influence of Taiwan on the track of a storm. The simulated winds at 2 km are easterly in the absence of the island. With the island present to block the flow, but without a hurricane present, the winds turn southerly to the east of Taiwan. With the hurricane present, the circulation around the storm, interacting with the simulated terrain, resulted in the path plotted in the Figure, which is to the right of the steering current in the absence of the storm. In the absence of the island, the hurricane would have moved on a general westward track.

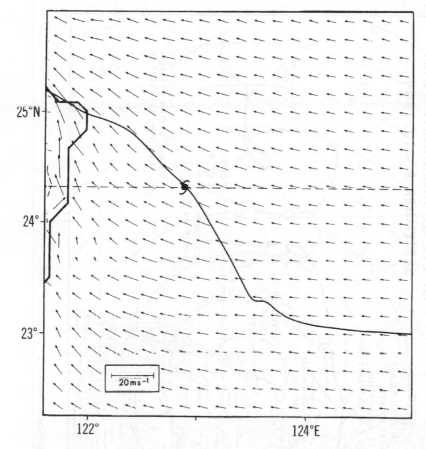

Figure 3.4 Numerical model simulation of the influence of the island of Taiwan on a hurricane track. Wind vectors at 2 km, owing to Taiwan and in the absence of the storm, are shown. Without the presence of Taiwan or the storm, the flow at 2 km is easterly.

Source: Bender *et al.* 1987

INTERNAL FLOW

Even when the steering current is uniform and steady, however, hurricane motion is often somewhat irregular around its mean motion. Figure 3.5 illustrates the trochoidal oscillation of the movement of the eye of Hurricane Dora (1964), as it progressed on a general track westward towards the upper east coast of Florida. In Hurricane Anita (1977), the storm centre was observed by aircraft to have a sinusoidal oscillation around the mean track with an amplitude of 5 km, and a period of 6 hours (Willoughby 1979). This small-scale irregular behaviour of the centre of the storm has been attributed to spatial asymmetries in the cumulus convection and frictional effects in the eye wall region, which causes the centre to deviate

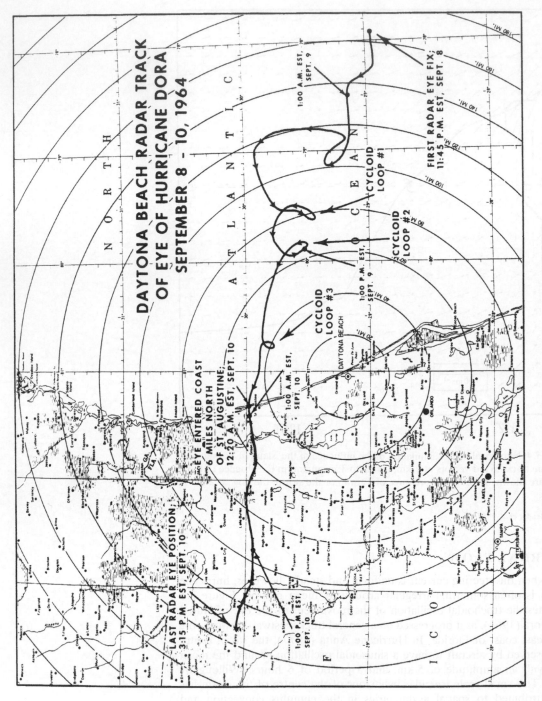

Figure 3.5 Daytona Beach radar track of eye of Hurricane Dora, 8–10 September 1964

Source: US Dept of Commerce 1964

periodically to the left or right of its track, similar to the motion of a spinning top. The larger general envelope of the storm, with its much greater inertia, more closely follows the steering current and tends to force the eye wall region back towards the centre of the larger circulation.

4 Impacts

The importance and danger of tropical cyclones differ between land and water. Over the oceans, the human activities at risk are primarily oil rigs, shipping, and air traffic. On land, particularly along the coast, cities, towns, and industrial activities become threatened.

OCEAN IMPACTS

The appearance of the ocean surface, as viewed from aircraft, for various wind speeds in three tropical cyclones is illustrated in Figures 4.1(a)–(gg). A verbal description of the effects of winds of various strengths over the ocean is presented in Table 4.1.

As evident in the Figures and Table, winds of hurricane speed over the ocean are characterized by blowing spray over a white, foaming sea state of large waves. Monstrous waves can develop in this environment. The expected relation between wave height, and fetch and length of time that the wind is at a given speed *with the same direction* (i.e. *the same fetch*) is presented in Tables 4.2(a)–(e). Of course, it is important to realize that, near the eye wall of a tropical cyclone, the wind direction varies significantly over short distances as air circulates around the storm. The result is that, while the waves may not reach the large heights possible for long fetches, the sea state will be chaotic, and an extreme hazard to shipping can occur in response to wave motion from a wide spectrum of directions.

As an example, the heights of the maximum ocean waves that were associated with Hurricane Kate in the eastern Gulf of Mexico in November 1985 are illustrated in Figures 4.2(a) and (b), along with other observed meteorological characteristics. These measurements were made from an unmanned ocean buoy. Wave height was determined by monitoring the length of time and magnitude of the vertical accelerations of the buoy by the waves. In this storm, a wave height of greater than 38 ft (11.6 m) was observed preceding the passage of the eye at 1800 GMT on 20 November.

Strong winds, of course, also occur in winter extratropical storms. The risk to shipping and other activities from wave action, however, is generally less serious in such storms because the wind direction, and hence the direction of propagation of swells and waves, is primarily in one direction in a given sector of the storm. The strongest winds in an extratropical cyclone also do not necessarily occur in its region of lowest pressure, since the pressure gradient force can be largest at a considerable distance from the centre. The magnitude of the pressure gradient is directly related to the strength of the near surface wind. The wave height estimates given in Tables 4.2(a)–(e) are most valid for extratropical storms.

In a hurricane, in contrast, the pressure field is concentric around the eye, with its greatest gradient in the eye wall region (e.g. see Figure 4.2(a)). Therefore, not only are the winds very strong in this environment but they are also changing direction rapidly around the eye. The result is a chaotic sea with swells and waves propagating in a myriad of directions. Estimates of wave height from tabulations such as those given in Tables 4.2(a)–(d) are thus not as useful in such an environment because of the greater likelihood that the large waves can superimpose on top of each other, producing enormous wave heights. In addition, since wind direction changes so rapidly around the eye, it is difficult to determine the appropriate fetch. A ship cannot simply steer into the running sea in order to reduce its risk since there is no one direction from which the waves come.

A second important difference between strong extratropical cyclones and hurricanes is that the winter cyclones seldom have speeds which exceed the minimal hurricane strength of 64 kn. In contrast, as already discussed in the Introduction, speeds greater than 155 kn. are occasionally observed in hurricanes. In 1944, in the western North Pacific during World War II, when an American naval convoy was inadvertently struck by a Pacific hurricane, three destroyers capsized and sank, nine other ships sustained serious damage, and 19 other vessels received lesser damage (Kotsch 1977).

LAND IMPACTS

At the coast and a short distance inland

At the coast, the major impacts of either a landfalling hurricane or one paralleling the coast are:

- storm surge,
- winds,
- rainfall,
- tornadoes.

Of these weather features, the storm surge accounts for over 90 per cent of the deaths in a hurricane.

STORM SURGE

'Storm surge' refers to a rapid rise of sea level that occurs as a storm approaches a coastline. This elevation of sea level is caused by (i) a lower overlying atmospheric pressure (because of the low pressure at the centre of a tropical cyclone, a 100 mb drop in ocean surface air pressure results in an elevation in the ocean of about 1 m); (ii) the piling up of water at the

60

Figure 4.1(a) Eloise (1975). Aircraft flight level: 5,276 ft (1,608 m); flight level winds: 6.9 ms⁻¹. Estimated one-minute averaged winds at 65 ft (19.8 m): 14 kn.

Source: Black and Adams 1983

Figure 4.1(b) Gladys (1975). Aircraft flight level: 515 ft (157 m); flight level winds: 7.7 ms⁻¹. Estimated one-minute averaged winds at 65 ft (19.8 m): 16 kn.

Source: Black and Adams 1983

Figure 4.1(c) Eloise (1975). Aircraft flight level: 1,542 ft (470 m); flight level winds: 8.3 ms⁻¹. Estimated one-minute averaged winds at 65 ft (19.8 m): 17 kn.

Source: Black and Adams 1983

Figure 4.1(d) Gladys (1975). Aircraft flight level: 509 ft (155 m); flight level winds: 8.6 ms⁻¹. Estimated one-minute averaged winds at 65 ft (19.8 m): 17 kn.

Source: Black and Adams 1983

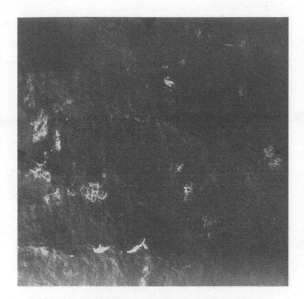

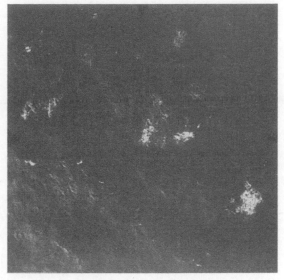

Figure 4.1(e) Gladys (1975). Aircraft flight level: 508 ft (155 m); flight level winds: 10.8 ms⁻¹. Estimated one-minute averaged winds at 65ft (19.8 m): 21 kn.

Source: Black and Adams 1983

Figure 4.1(f) Gladys (1975). Aircraft flight level: 502 ft (153 m); flight level winds: 11.1 ms⁻¹. Estimated one-minute averaged winds at 65 ft (19.8 m): 21 kn.

Source: Black and Adams 1983

Figure 4.1(g) Eloise (1975). Aircraft flight level: 794 ft (242 m); flight level winds: 15.5 ms⁻¹. Estimated one-minute averaged winds at 65 ft (19.8 m): 30 kn.

Source: Black and Adams 1983

Figure 4.1(h) Eloise (1975). Aircraft flight level: 863 ft (263 m); flight level winds: 16.6 ms⁻¹. Estimated one-minute averaged winds at 65 ft (19.8 m): 31 kn.

Source: Black and Adams 1983

62

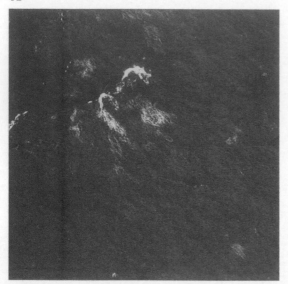

Figure 4.1(i) Eloise (1975). Aircraft flight level: 813 ft (248 m); flight level winds: 17.6 ms⁻¹. Estimated one minute averaged winds at 65 ft (19.8 m): 33 kn.

Source: Black and Adams 1983

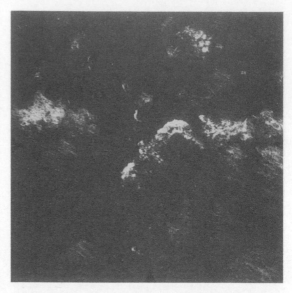

Figure 4.1(j) Eloise (1975). Aircraft flight level: 361 ft (110 m); flight level winds: 17.0 ms⁻¹. Estimated one-minute averaged winds at 65 ft (19.8 m): 33 kn.

Source: Black and Adams 1983

Figure 4.1(k) Gladys (1975). Aircraft flight level: 502 ft (153 m); flight level winds: 21.5 ms⁻¹. Estimated one-minute averaged winds at 65 ft (19.8 m): 41 kn.

Source: Black and Adams 1983

Figure 4.1(l) Eloise (1975). Aircraft flight level: 1,535 ft (468 m); flight level winds: 23.2 ms⁻¹. Estimated one-minute averaged winds at 65 ft (19.8 m): 43 kn.

Source: Black and Adams 1983

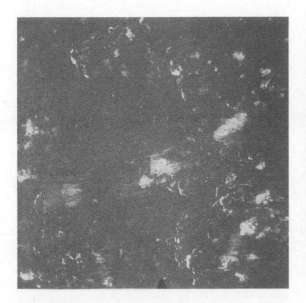

Figure 4.1(m) Eloise (1975). Aircraft flight level: 1,519 ft (463 m); flight level winds: 23.8 ms⁻¹. Estimated one-minute averaged winds at 65 ft (19.8 m): 44 kn.

Source: Black and Adams 1983

Figure 4.1(n) Eloise (1975). Aircraft flight level: 823 ft (251 m); flight level winds: 24.4 ms⁻¹. Estimated one-minute averaged winds at 65 ft (19.8 m): 45 kn.

Source: Black and Adams 1983

Figure 4.1(o) Eloise (1975). Aircraft flight level: 830 ft (253 m); flight level winds: 28.0 ms⁻¹. Estimated one-minute averaged winds at 65 ft (19.8 m): 53 kn.

Source: Black and Adams 1983

Figure 4.1(p) Eloise (1975). Aircraft flight level: 1,001 ft (305 m); flight level winds: 29.8 ms⁻¹. Estimated one-minute averaged winds at 65 ft (19.8 m): 54 kn.

Source: Black and Adams 1983

64

Figure 4.1(q) Gladys (1975). Aircraft flight level:
538 ft (164 m); flight level winds: 29.0 ms⁻¹. Estimated
one-minute averaged winds at 65 ft (19.8 m): 56 kn.

Source: Black and Adams 1983

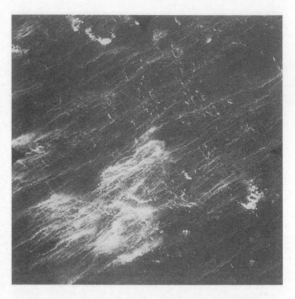

Figure 4.1(r) Eloise (1975). Aircraft flight level: 981 ft
(299 m); flight level winds: 32.0 ms⁻¹. Estimated
one-minute averaged winds at 65 ft (19.8 m): 58 kn.

Source: Black and Adams 1983

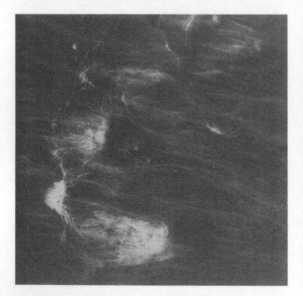

Figure 4.1(s) Gladys (1975). Aircraft flight level:
768 ft (234 m); flight level winds: 34.2 ms⁻¹. Estimated
one-minute averaged winds at 65 ft (19.8 m): 66 kn.

Source: Black and Adams 1983

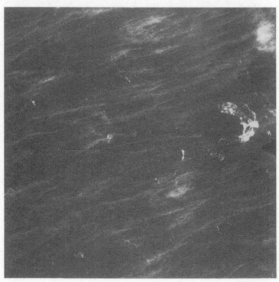

Figure 4.1(t) Eloise (1975). Aircraft flight level:
1,027 ft (313 m); flight level winds: 37.4 ms⁻¹.
Estimated one-minute averaged winds at 65 ft
(19.8 m): 68 kn.

Source: Black and Adams 1983

Figure 4.1(u) Eloise (1975). Aircraft flight level: 1,010 ft (308 m); flight level winds: 38.8 ms⁻¹. Estimated one-minute averaged winds at 65 ft (19.8 m): 71 kn.

Source: Black and Adams 1983

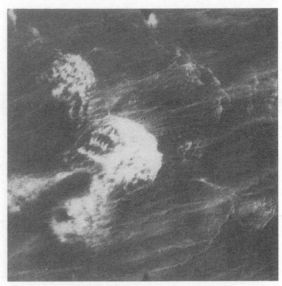

Figure 4.1(v) Gladys (1975). Aircraft flight level: 640 ft (195 m); flight level winds: 39.2 ms⁻¹. Estimated one-minute averaged winds at 65 ft (19.8 m): 73 kn.

Source: Black and Adams 1983

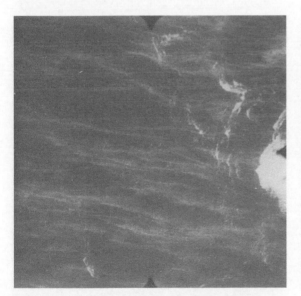

Figure 4.1(w) Eloise (1975). Aircraft flight level: 968 ft (295 m); flight level winds: 42.1 ms⁻¹. Estimated one-minute averaged winds at 65 ft (19.8 m): 77 kn.

Source: Black and Adams 1983

Figure 4.1(x) Eloise (1975). Aircraft flight level: 1,148 ft (350 m); flight level winds: 43.2 ms⁻¹. Estimated one-minute averaged winds at 65 ft (19.8 m): 78 kn.

Source: Black and Adams 1983

66

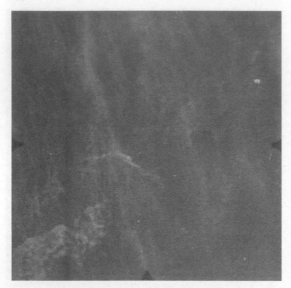

Figure 4.1(y) Gladys (1975). Aircraft flight level: 689 ft (210 m); flight level winds: 43.4 ms^{-1}. Estimated one-minute averaged winds at 65 ft (19.8 m): 82 kn.

Source: Black and Adams 1983

Figure 4.1(z) Eloise (1975). Aircraft flight level: 935 ft (285 m); flight level winds: 45.5 ms^{-1}. Estimated one-minute averaged winds at 65 ft (19.8 m): 83 kn.

Source: Black and Adams 1983

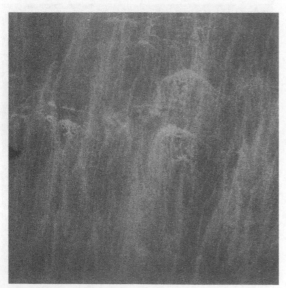

Figure 4.1(aa) Eloise (1975). Aircraft flight level: 991 ft (302 m); flight level winds: 49.1 ms^{-1}. Estimated one-minute averaged winds at 65 ft (19.8 m): 89 kn.

Source: Black and Adams 1983

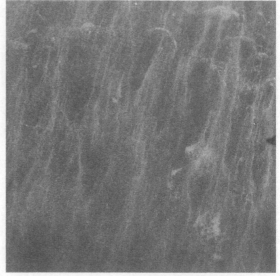

Figure 4.1(bb) Eloise (1975). Aircraft flight level: 1,004 ft (306 m); flight level winds: 49.4 ms^{-1}. Estimated one-minute averaged winds at 65 ft (19.8 m): 90 kn.

Source: Black and Adams 1983

Figure 4.1(cc) David (1979). Aircraft flight level: 1,643 ft (501 m); flight level winds: 58.8 ms⁻¹. Estimated one-minute averaged winds at 65 ft (19.8 m): 101 kn.

Source: Black and Adams 1983

Figure 4.1(dd) David (1979). Aircraft flight level: 1,647 ft (502 m); flight level winds: 58.8 ms⁻¹. Estimated one-minute averaged winds at 65 ft (19.8 m): 101 kn.

Source: Black and Adams 1983

Figure 4.1(ee) David (1979). Aircraft flight level: 1,486 ft (453 m); flight level winds: 61.7 ms⁻¹. Estimated one-minute averaged winds at 65 ft (19.8 m): 107 kn.

Source: Black and Adams 1983

Figure 4.1(ff) David (1979). Aircraft flight level: 1,469 ft (448 m); flight level winds: 61.7 ms⁻¹. Estimated one-minute averaged winds at 65 ft (19.8 m): 107 kn.

Source: Black and Adams 1983

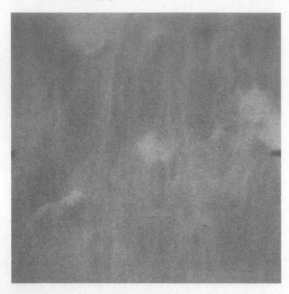

Figure 4.1(gg) David (1979). Aircraft flight level: 1,489 ft
(454 m); flight level winds: 63.0 ms⁻¹. Estimated
one-minute averaged winds at 65 ft (19.8 m): 110 kn.

Source: Black and Adams 1983

Table 4.1 The Beaufort wind scale for tropical cyclones from state-of-sea observations at 1,500 ft (457 m), except
for Beaufort numbers larger than 19 in which case observations are from 700 mb (about 10,000 ft/3.1 km)

Beaufort number	Descriptor	Estimated one-minute averaged wind at 65ft (19.8 m) in knots
0	Sea glassy. Appearance of being covered by oil.	Calm
1	Slight ripple.	2–6
2	Slight ripple. Isolated brief whitecaps. Unable to determine direction.	6–10
3	Surface like wrinkled paper. Small, well defined whitecaps of uniform size but few in number. White crests disappear quickly. First able to tell direction but with difficulty.	10–16
4	Small foam patches. Number of breaking crests increase slightly and are a little larger. First able to tell direction with confidence. Wrinkle texture of surface is very evident.	16–20
5	**Small craft warning.** Size and number of whitecaps and foam patches increase significantly. Whitecaps on most wave crests. Very short streaks may appear in foam patches.	20–26
6	Well defined short streaks in foam patches. Small whitecaps on most wave crests. Occasional medium-size foam patch or breaker. Isolated green patches of short duration. Foam patches, short streaks, and whitecaps (white water) cover 5–7% of sea surface.	26–32
7	Medium-size breaking crests. Dense foam patches and accompanying short	32–37

streaks are numerous. Average length of streaks equal to diameter of average foam patch. Small green patches occasionally visible.

8   **Gale warning–tropical storm.** Streaks more numerous and occasionally longer. Some streaks may appear unassociated with breaking waves or foam patches. Area covered by whitecaps stabilizes at 7–10%. Occasional large foam patch. Small green patches continually visible with occasional moderate-sized green patch.   37–44

9   Streaks readily apparent between foam patches. Streak length varies from patch size to occasional regions of long, nearly continual streaking. Streaks, patches, and breaking waves cover 15–20% of sea. 50% of foam patches are green.   44–50

10   **Storm warning.** Wind streaks become the most obvious surface feature and are continuously or nearly so. Well-defined, thinly breaking waves form on long crestlines, often preceded by short breaking wavelets giving a step-like appearance. Occasional large foam patches are quickly fragmented and elongated into streaks. Sea covered 20–25% by white water.   50–55

11   Streaks are well-defined, parallel, thin, close together, and continuous with very short capillary wavelets cutting across and perpendicular to streaks, giving sea surface a 'shattered glass' effect in certain areas. Some large breaking crests may take on 'rolling' or 'tumbling' appearance. Sea covered 30–40% by white water.   55–62

12   **Hurricane warning.** Sea may occasionally be obscured by spray and take on a murky appearance. Large, curved, breaking crests have undulating effect on streaks, giving churning appearance. Streaks appear to thicken and become milky or pale greenish.   62–69

13   Surface features generally become murky. Streaks and foam patches begin to lose their sharp definition and appear to smudge, thicken, or merge together. Frequent, extremely large, almost semicircular crests outlined by thinly breaking waves with occasional groups of large foam patches after entire crest breaks.   69–75

14   Quantity of spray increases. Streaks thicken and appear to have more depth. Previous crisp, shattered glass appearance now appears blurred. Most features appear to be a submerged rather than a surface phenomenon, owing to obscuration. Very short capillary wavelets which cut across streaks give surface a stressful appearance as though undergoing compression. Sea surface 50% white.   75–81

15   Sea appears flatter and entire surface takes on a whitish/greyish cast. Streaks organize somewhat into broader, diffuse bands. All features lose some definition and appear submerged. Surface 50–55% white.   81–88

16   Many thin streaks are partially obscured and those which can be seen may appear as bands spaced farther apart. Occasional cloud below aircraft blots out or obscures surface. Sea appears almost flat. Whitish cast covers 60–65% of surface.   88–95

17   Breaking waves and foam patches appear as diffuse, white, puffy areas. Streaks become fuzzy bands. Surface 70–80% white.   95–102

18   Cloud, spray, and foam patches merge into large, white, indefinable areas historically referred to as 'white sheets'. Surface features have only rough boundary definition.   102–108

19   Isolated large, white puffs. Only strongest features of previously seen thick streaks remain to be observed and result gives impression of only a very few widely scattered and non-parallel streaks or wide bands. Whitish and greenish cast covers 100% of surface.   108–115

Table 4.1 continued

20–21	Foam patches, bands, and whitecaps merge into large indefinable areas or white sheets. Variations in brightness are less distinct but still result in churning appearance.	115–129
22–23	Sea 100% white and green. Only slight variation in whiteness is apparent.	129–145

Source: Black and Adams 1983

Table 4.2 Relation between wind and wave characteristics

(a) Probable maximum heights of waves with various wind speeds and unlimited fetch

Wind speed (kn.)	Wave height (ft)
8	3
12	5
16	8
19	12
27	20
31	25
35	30
39	36
43	39
47	45
51	51

(d) Maximum wave heights with various wind speed and the fetches and durations required to produce waves 75 per cent as high as the maximum with each wind speed

Wind speed (kn.)	Max. wave height (ft)	75% of max. height (ft)	Fetch 75% (naut. miles)	Duration for 75% (hours)
10	2	1.5	13	5
20	9	6.8	36	8
30	19	14.3	70	11
40	34	25.5	140	16
50	51	38.3	200	18

(b) Wave heights (ft) produced by different wind speeds blowing over different fetches

Wind speed (kn.)	Fetch (nautical miles)					
	10	50	100	300	500	1,000
10	2	2	2	2	2	2
15	3	4	5	5	5	5
20	4	7	8	9	9	9
30	6	13	16	18	19	20
40	8	18	23	30	33	34
50	10	22	30	44	47	51

(e) Average wave length compared to wind speed

Average wave length (ft)	Wind speed (kn.)
52	11
124	20
261	30
383	42
827	56

Source: Kotsch 1977

(c) Wave heights (ft) produced by different wind speeds blowing for various lengths of time

Wind speed (kn.)	Duration (hours)						
	5	10	15	20	30	40	50
10	2	2	2	2	2	2	2
15	4	4	5	5	5	5	5
20	5	7	8	8	9	9	9
30	9	13	16	17	18	19	19
40	14	21	25	28	31	33	33
50	19	29	36	40	45	48	50
60	24	37	47	54	62	67	69

coast, generated by the strong onshore winds; and (iii) a decreased ocean depth on approaching the coast, which steepens the surge.

A storm surge is highest in the front right quadrant of a landfalling tropical cyclone, where the onshore winds are the strongest. It is also large where ocean bottom bathymetry focuses the wave energy (e.g. as in a narrowing embayment). Peak storm surge from a landfalling cyclone increases with lower central pressures and an increase in the radius of maximum winds out to 48 km (30 miles).

Storm surge also occurs when storms parallel the coast without making landfall. If a tropical cyclone moves along a coastline, such that onshore winds are to its rear with respect to its direction of motion, the storm surge will be larger than when the storm is moving ahead with the onshore winds in the direction of the cyclone movement. The storm surge will also lag the passage of the cyclone centre in the first case, while it will precede the storm centre when the cyclone is moving such that onshore winds are ahead of the storm. Offshore winds which are associated with a storm can produce a negative surge as the sea level is lowered by the strong winds blowing out from the coast.

Storm surge is estimated to diminish in depth by 1–2 ft (0.3–0.6 m) for every statute mile (1.6 km) that it moves inland. Even if the inland elevation were only 4–6 ft (1.2–1.8 m) above mean sea level, a storm surge of 20 ft (6.1 m) might reach no more than 7–10 statute miles (11–16 km) inland. Thus, the most destructive effect of the storm surge hazard is on beaches and offshore islands.

Storm surge calculations have been performed for much of the US coastline, using a computer model program called SPLASH (*Special Program to List Amplitudes of Surge from Hurricanes*) that was developed by C.P. Jelesnianski of the Systems Development Office of the National Weather Service (Jelesnianski 1974). For level 5 hurricanes (of the strength of the 1935 Florida Keys hurricane, Hurricane Camille [1969], and Hurricane Gilbert [1988]), SPLASH estimates, for example, a storm surge of 32 ft (9.7 m) on the coast near Cedar Key on the upper west coast of Florida. Figure 4.3 illustrates the estimated depth of ocean water penetrating inland south of Miami, owing to storm surge inundations of different intensities. The storm surge of 15 ft (4.6 m) corresponds to a level 5 hurricane. The storm surge estimates are obtained from SPLASH. Major damage and probably loss of life in this urbanized area would likely occur from storm surges greater than 5 ft (1.5 m). Waves and surf of several additional feet would be superimposed on top of the storm surge.

In 1900, 6,000 deaths occurred in Galveston, Texas primarily as a result of the storm surge that was associated with a Gulf of Mexico hurricane. In 1957, a storm surge, which was associated with Hurricane Audrey and which was over 12 ft (3.7 m) and extended as far inland as 25 statute miles (40 km), was the major cause of the death of 390 individuals in Louisiana. In September 1928, the waters of Lake Okeechobee, driven by hurricane winds, overflowed the banks of the lake and were the main cause of the 1836 deaths that were associated with the storm.

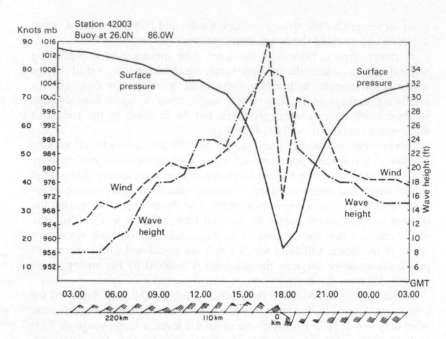

Figure 4.2(a) Observation of the passage of Hurricane Kate, 20 November 1985, as monitored by a floating oceanic buoy at 26° N and 86° W

Note: 1 ft = 0.3048 m

Source: Plotted by R.H. Johnson 1985, unpublished

WINDS

The strong winds of a hurricane can produce considerable structural damage and risk to life from flying debris, even inland from the coast. While winds reduce after landfall, as the pressure gradient of the storm lessens, destructive winds can still occur far inland.

The damage from winds is proportional to the kinetic energy of the flow; thus, a wind of 50 ms^{-1} is four times as effective at causing damage as a wind of 25 ms^{-1}. This relation between wind speed and kinetic energy is shown in Figure 4.4.

Maximum gusts, of course, are even stronger than reported one-minute average sustained winds. Although the former were not reported in earlier years, an estimate of maximum likely gusts can be obtained from:

$$V_{\text{max gust}} = V + 2.15\,\sigma$$

where σ is the standard deviation of the wind and V is the average wind over a time period of around 20 minutes. An estimate for σ based on theoretical analysis of turbulence near the surface is:

$$\sigma = \frac{0.8\,V}{\ln\,(z/z_0)} \tag{4.1}$$

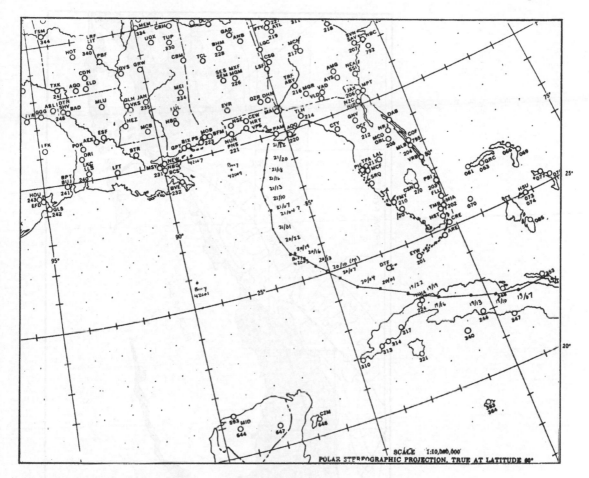

Figure 4.2(b) Track of Hurricane Kate, 19–22 November 1984

Source: Plotted by R.H. Johnson 1985, unpublished

where z is the height at which the average wind was observed. The height z_0, which corresponds to the level where the average wind becomes zero, is defined as:

$$z_0 = \frac{0.006}{g} \, \sigma^2 \qquad (4.2)$$

where g is the acceleration of gravity. Since σ occurs in both equations 4.1 and 4.2, iteration must be used to obtain a value of σ which satisfies both equations. As an example, for an average wind of $50 \, ms^{-1}$ at a height of 5 m, $\sigma = 8.5 \, ms^{-1}$ with $z_0 = 4$ cm. Thus, an estimate of the peak gust is $68 \, ms^{-1}$.

74

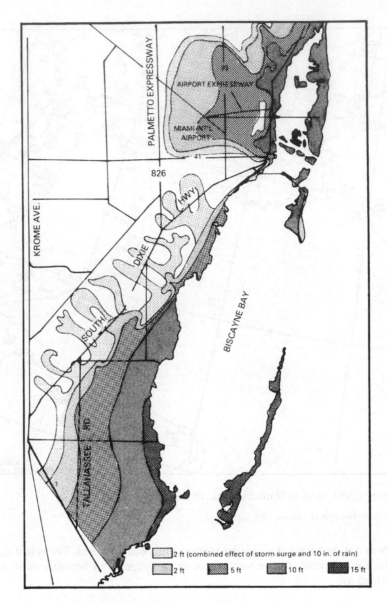

Figure 4.3 Estimated storm surge owing to a level 5 hurricane landfalling south of Miami, Florida.

Note: 1 ft = 0.3048 m; 1 in. = 25.4 mm

Source: US Dept of Commerce 1977

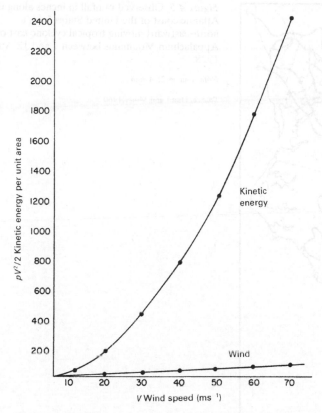

Figure 4.4 Relation between wind speed and kinetic energy. A density of 1 kg m^{-3} was assumed for this example.

RAINFALL

Rainfall is often excessive at and after a tropical cyclone makes landfall, particularly if the very moist air of the storm is forced up and over mountain barriers. Figure 4.5 illustrates observed rainfall associated with an August Atlantic tropical cyclone of 1928 as it moved up the Atlantic coast east of the Appalachians. Even relatively weak tropical-like disturbances can result in extreme rainfall, as seen, for example, over coastal Texas in September 1979 in which upwards of 19 in. (483 mm) of rain inundated the area over a period of several days (Bosart 1984). Occasionally, for reasons not completely understood, rainfall is light in the vicinity of hurricanes. Hurricane Inez in 1966, for instance, resulted in only a few drops of rain in Miami for several hours when the centre was south and south-southwest of Miami and at its closest point to the city. At the time, Miami was under the wall cloud and, normally, torrential rains would have been expected. As a result of the absence of rain, the strong winds advected salt spray many kilometres inland, causing severe damage to vegetation from salt accumu-

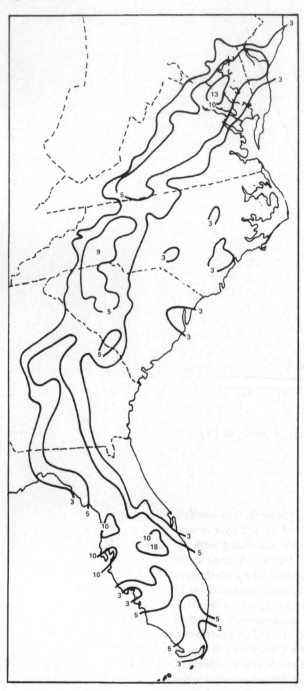

Figure 4.5 Observed rainfall in inches along the Atlantic coast of the United States from a north-eastward moving tropical cyclone east of the Appalachian Mountains between 7 and 12 August 1928.

Note: 1 in. = 25.4 mm

Source: Dunn and Miller 1960

lation. Homestead Air Force Base, south of Miami and closer to the hurricane centre, received only 0.62 in. (16 mm) of rain during the entire storm.

TORNADOES

Tornadoes are the fourth threat from tropical cyclones near the coast. These rapidly rotating small-scale vortices are spawned in squalls, usually in the front right quadrant of the storm. It is felt the tornadoes develop in response to the large vertical shears of the horizontal wind that develop as the lower level winds are retarded by ground friction. A tilting of this large velocity shear into the horizontal plane by spatially varying vertical motion, which occurs in the hurricane squall line environment, provides the circulation needed for tornadogenesis (see the schematic in Figure 4.6). While these tornadoes are not often as severe as the major tornadoes that are associated with spring-time continental convective storm systems, loss of life and property damage does occur as a result of them. Figure 4.7(a)

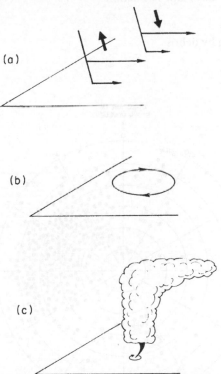

(a)

(b)

(c)

Figure 4.6 Schematic of a postulated mechanism for tornadogenesis in a hurricane environment. (a) Tilting of strong vertical shear of the horizontal wind by cumulus convection in one location (ascent) and compensating subsidence adjacent to the cumulus updraft. (b) Resultant generation of horizontal eddy as vertical shear of the horizontal wind is tilted to some extent into the horizontal plane. (c) Development of subsequent cumulus convection over the eddy concentrates and speeds up the horizontal circulation until it becomes a tornado.

shows observed locations of tornadoes with respect to the orientation of landfalling storms. Occasionally, landfalling storms produce tornado swarms, such as the tornadoes that were observed to be associated with the landfall of Hurricane Beulah (1969) on the Texas coast, as illustrated in Figure 4.7(b). Hurricanes Carla (1961) and Celia (1970), in contrast, produced no observed tornadoes.

COASTAL ZONING

In Texas, the Texas Coastal and Marine Council has identified four zones that reflect levels of exposure to several of the hurricane threats. Figures 4.8(a) and (b) illustrate the four zones defined as:

Zone A Wind
        Flooding
        Battering by debris
        Scour by wave and surf action
Zone B Wind
        Flooding
        Battering by debris
Zone C Wind
        Flooding
Zone D Wind

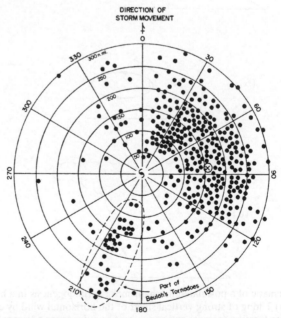

Figure 4.7(a) Tornado occurrence in the United States (1948–72) with respect to direction of movement of landfalling hurricanes. The symbol ⊗ is the centroid of the observed tornado distribution.

Source: Novlan and Gray 1974

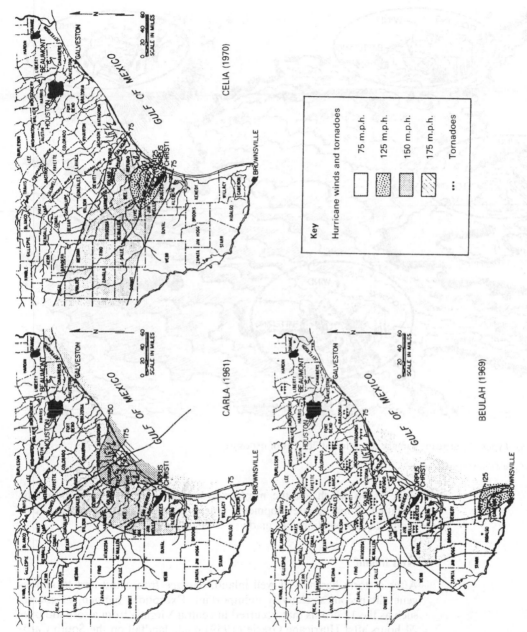

Figure 4.7(b) The track of the eyes of Hurricanes Carla, Beulah, and Celia, and the area in Texas covered by hurricane-force winds (in statute m.p.h.)

Note: 1 statute m.p.h. = 0.87 kn.

Source: Texas Coastal and Marine Council 1974

Figure 4.8(a) Types of hurricane damage for different degrees of exposure

Source: Texas Coastal and Marine Council 1976

Zone A, in which are the most exposed locations, probably should be excluded from all development unless stringent, and perhaps cost-prohibitive, protections are made.

Well inland

After a storm has moved well inland, damage from wind and tornadoes generally becomes relatively unimportant. Exceptions happen, of course, such as the 12 deaths that occurred in central Virginia from a tornado over 24 hours after Hurricane Gracie (1959) made landfall on the South Carolina coast.

The biggest threat to life and property occurs as a result of flash flooding and riverine flooding from excessive rainfall. Particularly dangerous are tropical cyclones whose rainfall becomes light and benign after landfall only to erupt a couple of days later into torrential downpours when the

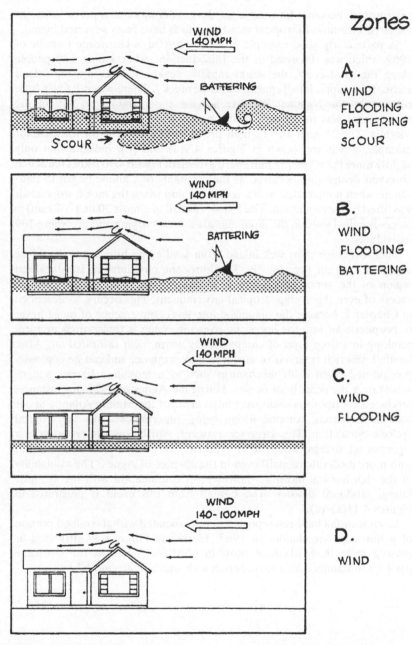

Zones

A.
WIND
FLOODING
BATTERING
SCOUR

B.
WIND
FLOODING
BATTERING

C.
WIND
FLOODING

D.
WIND

Figure 4.8(b) Schematic representation of hazard zones A to D in Texas coastal areas

Note: 1 statute m.p.h. = 0.87 kn.

Source: Texas Coastal and Marine Council 1976

environment becomes favourable for the condensation and precipitation of the large quantities of tropical moisture which have been advected inland.

A particularly good example of such a system is Hurricane Camille of 1969 which was discussed in the Introduction. After killing 139 people along the Gulf coast, the storm rapidly weakened after moving inland across Mississippi, into Tennessee and Kentucky. There was relatively little concern by the National Weather Service and certainly no hint of the tragedy that was to happen on the night of 19 August 1969 in central Virginia. The 24-hour and 12-hour precipitation forecasts for the area, for example, which are shown in Figures 4.9(a) and (b), indicate that only slightly more than 2 in. (51 mm) were expected. Figure 4.10 shows the actual observed deluge that occurred as the remnants of Camille began to rejuvenate when it interacted with a cold front and when the moist, tropical air was lifted by the mountains. The 6-hour rainfall of almost 30 in. (762 mm) in places liquified soils on the mountainous slopes, burying and drowning 109 individuals.

Such excessive rains well inland from landfalling hurricanes should be expected to occur occasionally. The hurricane environment is a localized region of the atmosphere which is enriched with water vapour, well in excess of even the average tropical environment. This occurs, as described in Chapter 2, because the organized low-level convergence of moist lower tropospheric air into the hurricane over the ocean is transported upward, resulting in a deep layer of comparatively warm, near saturated air. After landfall, this rich reservoir of moisture moves inland, and can be copiously precipitated when a lift mechanism such as a mountain barrier and/or ascent over the polar front occurs. Hurricane Agnes of 1972, for instance, produced enormous rainfalls over large areas of the Middle Atlantic States because of strong synoptic lifting being superimposed on the tropical cyclone circulation. This large-scale ascent, which was associated with a vigorous jet stream, would have produced an extratropical cyclone (but with more moderate rainfall) even in the absence of Agnes. The availability of the rich tropical moisture, however, in combination with the synoptic lifting, produced disaster. The rainfall from this event is illustrated in Figures 4.11(a)–(c).

Even snowfall has been reported to be associated with the inland portion of a hurricane circulation. In 1963, Hurricane Ginny was attributed as causing more than 30 cm of snow in northern Maine as the hurricane passed and slammed into Nova Scotia with winds of around 87 kn.

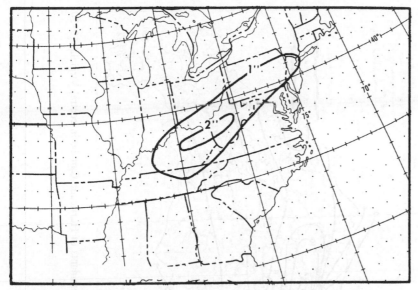

Figure 4.9(a) Forecast precipitation in inches for 24 hours ending 1200 GMT,
20 August 1969

Note: 1 in. = 25.4 mm

Source: US Dept of Commerce 1969

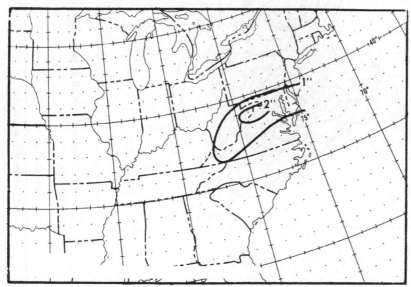

Figure 4.9(b) Forecast precipitation in inches for 12 hours ending 1200 GMT,
20 August 1969

Note: 1 in. = 25.4 mm

Source: US Dept of Commerce 1969

84

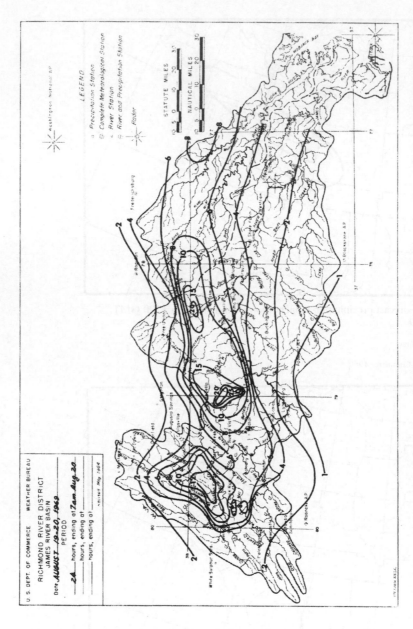

Figure 4.10 Observed rainfall in inches associated with the reminants of Hurricane Camille in central Virginia, 19–20 August 1969.

Note: 1 in. = 25.4 mm

Source: US Dept of Commerce 1969

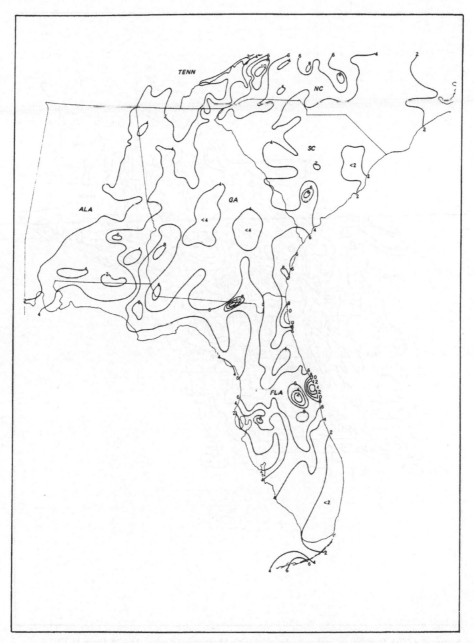

Figure 4.11(a) Rainfall in inches associated with Hurricane Agnes, 18–25 June 1972

Note: 1 in. = 25.4 mm

Source: DeAngelis and Hodge 1972

86

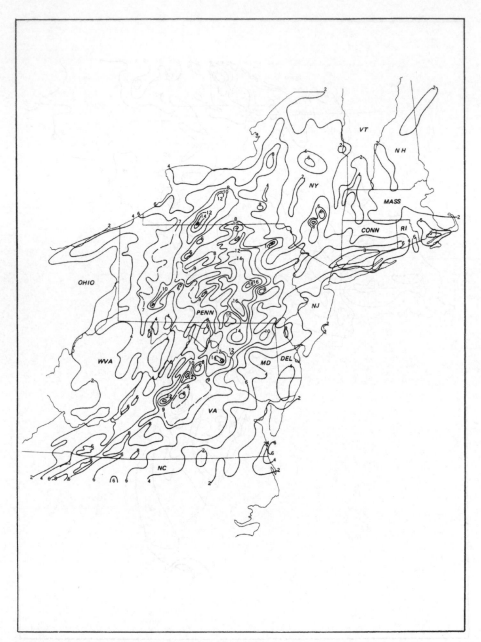

Figure 4.11(b) Rainfall in inches associated with Hurricane Agnes, 18–25 June 1972

Note: 1 in. = 25.4 mm

Source: DeAngelis and Hodge 1972

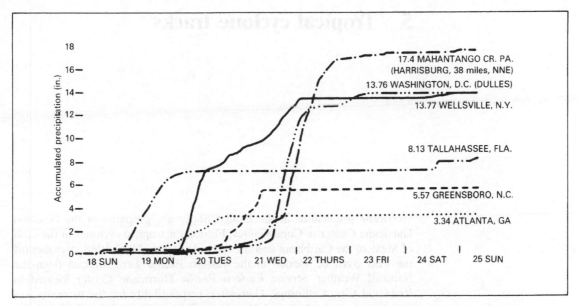

Figure 4.11(c) Cumulative rainfall curves in inches for selected locations during Hurricane Agnes, 18–25 June 1972

Note: 1 in. = 25.4 mm

Source: US Dept of Commerce 1973

5 Tropical cyclone tracks

Hurricane predictions in the United States are prepared at the National Hurricane Center in Coral Gables, Florida for tropical cyclones in the Gulf of Mexico, the Caribbean and the Atlantic Ocean. Until 1988, cyclones off the west coast of Mexico in the eastern Pacific were forecast from the National Weather Service Eastern Pacific Hurricane Center located in Redwood City, California. Thereafter, responsibility for the forecasts was transferred to the National Hurricane Center. Central Pacific hurricane predictions emanate from Honolulu, Hawaii.

In order to help the public and interested agencies to keep track of specific cyclones, lists are prepared of names to be used. Table 5.1 presents the names to be used for 1990–1 for the Atlantic Ocean, Gulf of Mexico

Table 5.1 Atlantic hurricane names for 1990 and 1991

1990	*1991*
Arthur	Ana
Bertha	Bob
Cesar	Claudette
Diana	Danny
Edouard	Elena
Fran	Fabian
Gustav	Gloria
Horrtense	Henri
Isidore	Isabel
Josephine	Juan
Klaus	Kate
Lili	Larry
Marco	Mindy
Nana	Nicholas
Omar	Odette
Paloma	Peter
Rene	Rose
Sally	Sam
Teddy	Teresa
Vicky	Victor
Wilfred	Wanda

and Caribbean Sea. A map, prepared by NOAA, which can be used to track tropical cyclones is presented in Appendix B.

TROPICAL CYCLONE TRACK PREDICTIONS

The National Hurricane Center applies several track models to forecast the movement of tropical cyclones. Figure 5.1 presents an example of the prediction of the movement of Hurricane Frederic (1979) from its initial position, at 7 a.m. on 11 September, north of the coast of Cuba. The models listed as NHC 67, NHC 72, and NHC 73 are statistical models which relate storm motion to large-scale observed and predicted atmospheric conditions. Statistical analysis of prior tropical cyclones permits the development of expected storm movement which is based on a given set of current and forecast synoptic-scale meteorological flows. The three models utilize different sets of statistical predictors.

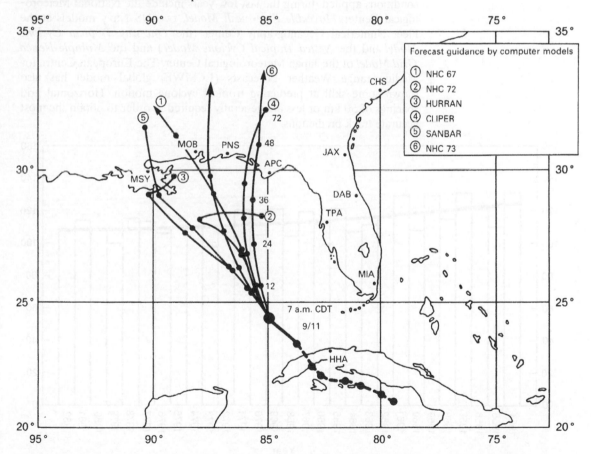

Figure 5.1 Forecast positions, generated by six computer models, and the official forecast track for Hurricane Frederic, 7 am. CDT, Tuesday, 11 September 1979

Source: Carter 1983

The HURRAN model is an analogue model (*Hurri*cane *An*alogue) which selects the most common subsequent tracks of a storm, located at a specific latitude and longitude at a given date, based on all previous storms which were similarly located. The CLIPER model (*Cli*matology-*Per*sistence) determines the expected motion of a storm, at a given location, based on a combination of the climatologically-expected movement and the current observed direction and speed of motion (persistence).

The SANBAR model (*San*ders *Bar*otropic, named after the originator of the model, Professor F. Sanders of MIT) solves a simplified set of physical equations which describe the movement of a vortex in a specified large-scale flow. The large-scale flow is determined from the observed atmospheric conditions at any given time.

More physically complete three-dimensional numerical models of tropical cyclone motion using observed and larger scale modelled atmospheric conditions applied during the last few years include the National Meteorological Center *Moveable Fine-mesh Model*, two US Navy models at the Fleet Numerical Oceanography Center (the *One-way Tropical Cyclone Model* and the *Nested Tropical Cyclone Model*) and the *Multiple-Nested Grid Model* of the Japan Meteorological Center. The European Centre for Medium-range Weather Forecasts (ECMWF) global model has also shown some skill at predicting tropical cyclone motion. Horizontal grid spacings of 50 km or less are generally required in order to obtain the most accurate track predictions.

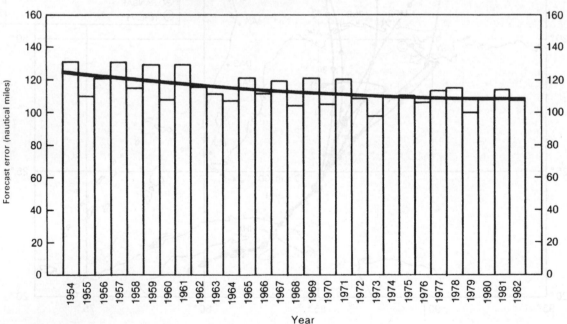

Figure 5.2 Average error of 24-hour forecast positions, in nautical miles, 1954–82

Note: 1 nautical mile = 1.83 km

Source: Carter 1983

These models are used to obtain the best estimate of tropical cyclone motion. Figure 5.2 illustrates the trend and accuracy of 24-hour forecasts of storm position between 1954–82. Note that an improvement of only 15 nautical miles (27.5 km) has been achieved, despite the great advances both in monitoring these storms (e.g. radar, satellite, reconnaissance aircraft) and in computer power to process and analyse the data.

Figure 5.3 illustrates the practical implication of the error in forecast position. Over a period of 48 hours, the predicted landfall of Hurricane Frederic (1979) varied from near Biloxi, Mississippi to the west-central panhandle of Florida. The actual landfall was just east of Mobile, Alabama. Causing $752,500,000 of damage, the storm was one of the most destructive in terms of property losses in US history up to that time.

An example of a verbal hurricane advisory bulletin, which was disseminated by the National Hurricane Center, was presented in the Introduction. Since 1983, probabilities of a tropical cyclone passing within 60 nautical miles (110 km) of specific geographic locations have also been publicly distributed. An example of the format used in these probability forecasts is shown in Figure 5.4, in this case for Hurricane Frederic.

Tropical cyclone intensity change predictions

Forecasts of tropical cyclone intensity change and development generally rely on a decision-tree approach in which satellite and atmospheric analyses are used to estimate whether or not conditions are favourable for a change. Figures 5.5(a) and (b) present an example of the type of decision tree that is used by the National Hurricane Center. In practice, the National Hurricane Center becomes concerned, regarding tropical cyclone development during the climatological season, whenever a south-west surface wind is observed south of 30°N in the Atlantic or eastern Pacific northern hemisphere.

Tropical cyclone-related public forecasts

The National Hurricane Center issues the following types of specific advisory bulletins:

- *Hurricane Watch*: Issued for a coastal area when there is a threat of hurricane conditions within 24–36 hours.
- *Hurricane Warning*: Issued when hurricane conditions are expected in a specified coastal area within 24 hours or less. Hurricane conditions include winds of 74 m.p.h. (64 kn.) and/or dangerously high tides and waves. Actions for protection of life and property should begin immediately when the warning is issued.
- *Small Craft Cautionary Statements*: When a tropical cyclone threatens a coastal area, small craft operators are advised to remain in port or not to venture into the open sea.
- *Gale Warning*: May be issued when winds of 39–54 m.p.h. (34–47 kn.) are expected.
- *Storm Warning*: May be issued when winds of 55–73 m.p.h. (48–63 kn.) are expected. If a hurricane is expected to strike a coastal area, gale or storm warnings will not usually precede hurricane warnings.

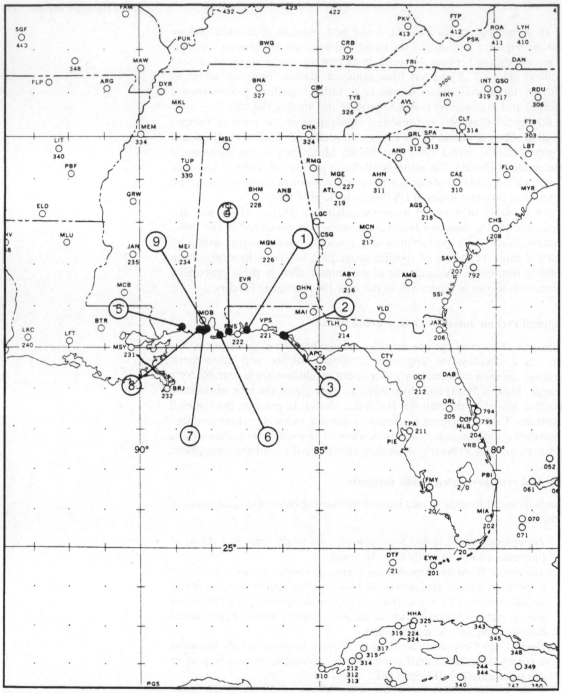

Figure 5.3 Successive predicted landfall locations for Hurricane Frederic from 1 p.m. CDT, Monday, 10 September to 1 p.m. CDT, Wednesday, 12 September 1979

Source: Carter 1983

PUBLIC ADVISORY # 52 ISSUED AT 9:30 PM CDT TUE SEPT 11 1979

HURRICANE FREDERIC PROBABILITIES
FOR GUIDANCE IN HURRICANE PROTECTION PLANNING
BY GOVERNMENT AND DISASTER OFFICIALS

CHANCES OF CENTER OF FREDERIC PASSING WITHIN 65 MILES OF
LISTED LOCATIONS THROUGH 7 PM CDT FRIDAY SEPTEMBER 14 1979

CHANCES EXPRESSED IN PER CENT . . . TIMES CDT

	THRU 7 PM WED	ADDITIONAL INCREMENTS 7 PM WED THRU 7 AM THU	7 AM THU THRU 7 PM THU	7 PM THU THRU 7 PM FRI	TOTAL THRU 7 PM FRI
COASTAL LOCATIONS					
MARCO ISLAND, FL	—	—	—	1	1
FT. MYERS, FL	—	1	—	—	1
VENICE, FL	1	—	1	—	2
TAMPA, FL	1	1	1	1	4
CEDAR KEY, FL	2	3	1	1	7
ST. MARKS, FL	7	5	2	—	14
APALACHICOLA, FL	16	3	—	1	20
PANAMA CITY, FL	19	3	—	1	23
PENSACOLA, FL	21	3	1	—	25
MOBILE, AL	16	6	1	—	23
GULFPORT, MS	14	6	1	1	22
BURAS, LA	16	4	1	—	21
NEW ORLEANS, LA	8	7	1	1	17
NEW IBERIA, LA	1	6	3	2	12
PORT ARTHUR, TX	—	1	3	3	7
GALVESTON, TX	—	1	2	2	5
PORT O'CONNOR, TX	—	—	1	2	3
CORPUS CHRISTI, TX	—	—	1	1	2
BROWNSVILLE, TX	—	—	—	1	1

— PROBABILITY LESS THAN 1 PER CENT

Figure 5.4 Probabilities of Hurricane Frederic being within 60 nautical miles
(110 km) of selected geographical locations at selected periods up to 72 hours from
the time of the advisory bulletin

Source: Carter 1983

Local National Weather Service Offices will include this information in
their public, aviation, and marine forecasts and will also disseminate their
own warnings such as special marine warnings. The Severe Storms Forecast
Center in Kansas City will also disseminate the advisory bulletins listed
below when a threat of tornadoes or severe thunderstorms is perceived.

• *Tornado Watch*: Conditions are favourable for the development of
severe thunderstorms which contain tornadoes.

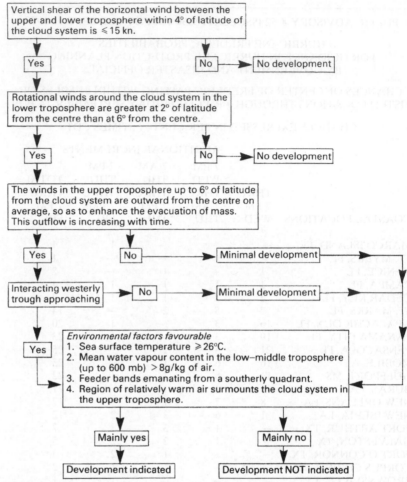

Figure 5.5(a) Decision tree used to estimate whether a tropical cyclone will develop

Source: Adapted from Simpson 1971

- *Severe Thunderstorm Watch*: Conditions are favourable for the development of thunderstorms with frequent lightning accompanied by straight-line damaging winds of greater than 60 kn. and hail greater than ³/₄ in. (1.9 cm) in size.

Local National Weather Service offices will issue Tornado Warnings and Severe Thunderstorm Warnings when radar or visual observations indicate that these weather events are actually occurring.

Flash flood and riverine flood information is distributed by National Weather Service River Forecast Centers and River District Offices. Official definitions of the watches and warnings include the following.

- *Flash Flood Watch*: Means a flash flood is possible in the area; stay alert. Flash Flood Watches are issued by River District Offices.

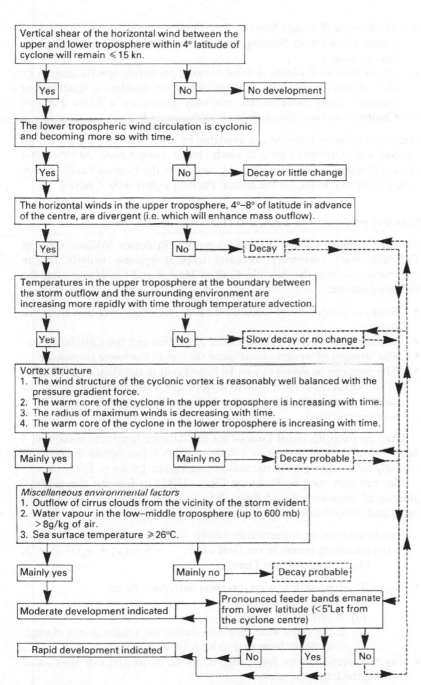

Figure 5.5(b) Decision tree used to estimate whether an existing tropical cyclone will intensify. Note that the presence of a poleward spiralling feeder band is weighted heavily in the intensification potential. The dashed lines represent the check of this condition for each of the dashed boxes.

Source: Adapted from Simpson 1971

- *Flash Flood Warning*: Means a flash flood is imminent; take immediate action. Flash Flood Warnings can be issued by any National Weather Service facility.
- *Flood Forecast Bulletin*: A flood bulletin, predicting specific stages of a river at specific locations, is issued whenever flooding is imminent or existing. These Bulletins are normally issued by a River Forecast Center.

Hurricane advisory bulletins are available by telephone to the public from the National Hurricane Center in south Florida, from 1 June–30 November in the Atlantic (1–900–410–NOAA), and from the Central Pacific Hurricane Center in Hawaii, for the central Pacific (1–900–410–CANE).

Seasonal predictions of tropical cyclone activity

Trends for entire seasons are also prepared. Professor William Gray of Colorado State University forecasts tropical cyclone activity for an upcoming season in the Atlantic, Gulf of Mexico and Caribbean using the following criteria:

- Status of winds 65,000–100,000 ft (19.8–30.5 km) over the equatorial region.
- Sea-level air pressure over the Gulf of Mexico and the Caribbean Sea.
- The strength of westerly winds near the top of the lower troposphere.
- The presence or absence of an El Niño event in the Pacific (an El Niño is an occasional radical shift to abnormally warm temperatures in the eastern tropical Pacific Ocean which results in world-wide shifts in weather patterns).

Gray prepares his initial forecast for the Atlantic hurricane season on 1 June, followed by an update on 1 August which is just before the climatologically most active part of the season (see Figure 1.6 on p. 19).

The equation used by Professor Gray (1983) to forecast the seasonal number of hurricanes is as follows. The seasonal forecast for both hurricanes and tropical storms has a similar form.

$$\text{predicted number of hurricanes during the upcoming season in the Gulf of Mexico and Atlantic Ocean} = 6 + (a_1 + a_2) + a_3 + a_4$$

- $a_1 = 30$ mb equatorial wind direction correction factor.
    - (i) If westerly, $a_1 = 1$,
    - (ii) If easterly, $a_1 = -1$,
    - (iii) If the zonal wind direction during the season is in a change-over phase from east to west, or west to east, $a_1 = 0$.
- $a_2 =$ correction factor for change in 30 mb equatorial east-west winds during the hurricane season.
    - (i) If uniformly increasing westerly with time, $a_2 = 1$,
    - (ii) If uniformly decreasing westerly with time, $a_2 = -1$,
    - (iii) If the east–west wind direction reverses from either

uniformly increasing westerlies or uniformly decreasing westerlies during the season, $a_2 = 0$.

The variable $a_1 + a_2$ represents the quasi-biennial oscillation (QBO) effect on hurricane activity. The QBO is discussed briefly on p. 23.

- $a_3 =$ the El Niño effect.
    - (i) For a moderate El Niño, $a_3 = -2$,
    - (ii) For a strong El Niño, $a_3 = -4$,
    - (iii) Otherwise, $a_3 = 0$.

An El Niño occurs when the normal flow equatorward and the resultant ocean upwelling of cold water along the west coast of northern South America is disrupted, resulting in anomalously warm temperatures in the equatorial Pacific off the South American coast. Apparently, an alteration in weather patterns in this region is statistically associated with changes in the atmosphere over the Gulf of Mexico and Atlantic Ocean so as to provide a more favourable environment for hurricanes.

- $a_4 =$ average sea level pressure deviation from the April–May or June–July climatological mean in the Caribbean Basin (six meteorological stations are used to obtain this measure: Brownsville, Texas; Merida, Mexico; Miami, Florida; San Juan, Puerto Rico; Curaçao, Venezuela; Barbados).

    - (i) If the pressure anomaly is between −0.4 mb and −0.8 mb, $a_4 = 1$,
    - (ii) If the pressure anomaly is less than −0.8 mb, $a_4 = 2$,
    - (iii) If the pressure anomaly is between 0.4 mb and 0.8 mb, $a_4 = -1$,
    - (iv) If the pressure anomaly is greater than 0.8 mb, $a_4 = -2$,
    - (v) If the pressure anomaly is between −0.4 mb and 0.4 mb, $a_4 = 0$.

Gray cautions that, in the use of this equation, if the equation indicates a value less than three during an El Niño year, then disregard and made a seasonal forecast of at least three hurricanes. If the equation indicates a value less than four for a non-El Niño year, then disregard and make the forecast for four hurricanes.

For the years 1984–9, Gray made the predictions listed in Table 5.2. The observed occurrences are also presented in the Table. Although five years of data are too few to be definitive regarding the skill of Gray's long-range forecast technique, the results so far are very encouraging even with the poor forecast year of 1989, which Gray attributes to the absence of a predictor in his seasonal forecast equation based on rainfall patterns over the Sahel in Africa. Years with normal or above average rainfall in this part of Africa appear to be related to an increase in the number of tropical cyclones in the Atlantic which develop as the disturbances move westward off the African coast.

ATTEMPTS AT TROPICAL CYCLONE MODIFICATION

There have also been attempts to modify tropical cyclones. The main hypothesis is that by seeding cumulus clouds with silver iodide outside the eye

Table 5.2 The prediction of seasonal tropical cyclone activity in the Atlantic Ocean, Gulf of Mexico and Caribbean Sea (1984–9) and observed occurrences

1984	Prediction as of 29 May	Updated prediction of 30 July	Observed
No. of hurricanes	7	7	5
No. of hurricane days	30	30	21
No. of hurricane and tropical storms	10	10	12
No. of hurricane and tropical storm days	45	45	61
		(implied from hurricane forecast)	

1985	Prediction as of 28 May	Updated prediction of 27 July	Observed
No. of hurricanes	8	7	7
No. of hurricane days	35	30	29
No. of hurricane and tropical storms	11	10	11
No. of hurricane and tropical storm days	55	50	60
		(implied from hurricane forecast)	

1986	Prediction as of 29 May	Updated prediction of 28 July	Observed
No. of hurricanes	4	4	4
No. of hurricane days	15	10	13
No. of hurricane and tropical storms	8	7	6
No. of hurricane and tropical storm days	35	25	27
		(implied from hurricane forecast)	

1987	Prediction as of 28 May	Updated prediction of 27 July	Observed
No. of hurricanes	5	4	3
No. of hurricane days	20	15	7
No. of hurricane and tropical storms	8	7	7
No. of hurricane and tropical storm days	40	35	36
		(implied from hurricane forecast)	

1988	Prediction as of 29 May	Updated prediction of 28 July	Observed
No. of hurricanes	7	7	5
No. of hurricane days	30	30	26
No. of hurricane and tropical storms	11	11	12
No. of hurricane and tropical storm days	50	50	56
		(implied from hurricane forecast)	

1989	Prediction as of late May	Updated prediction of 1 Aug	Observed
No. of hurricanes	4	4	7
No. of hurricane days	15	15	35
No. of hurricane and tropical storms	7	9	10
No. of hurricane and tropical storm days	30	35	66
		(implied from hurricane forecast)	

wall, clouds with liquid water colder than 0°C (referred to as *supercooled water*) could be converted to ice crystals. This change of phase of water would release heat of fusion, thereby enhancing the growth of the cumulus clouds and establishing the eye wall at a greater radius from the centre of the storm circulation. Without the silver iodide, it is hypothesized that the cloud droplets would remain liquid. Just as ice skaters slow their rotation when their arms are spread out, the hypothesis is that a larger radius of the eye wall will cause a reduction in wind strength. The hypothesis is sketched schematically in Figure 5.6.

This programme of tropical storm modification has been called Project Stormfury. Robert and Joanne Simpson were the original source of the Stormfury hypothesis in 1960. Their insight was inspired by an observation of Hurricane Donna (1960) by Professor Herb Riehl, who noted that nearly all of the outflow cloudiness stemmed from an aggregation of cumulus convection in the front right quadrant of the eye wall. Hurricanes Esther (1961), Beulah (1963), and Debbie (1969) have been seeded as part of this project, although only the Debbie experiments closely followed the current Stormfury hypotheses.

In an earlier experiment, a hurricane was seeded on 13 October 1947 off the south-east United States coast. Because it subsequently moved westward into the Georgia coast, questions were raised among critics of weather modification as to whether the seeding caused the abrupt change

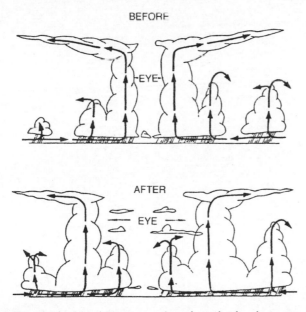

Figure 5.6 Hypothesized vertical cross-sections through a hurricane eye wall and rain bands before and after seeding. Dynamic growth of seeded clouds in the inner rain bands provides new conduits for conducting mass to the outflow layer and causes decay of the old eye wall.

Source: Simpson *et al.* 1978

in storm track. More recent analysis strongly suggests that the alteration in direction would have occurred in any case. Hurricane Ginger was also seeded in 1971. However, it was an anomalous storm with an eye wall usually below 6 km and no significant quantities of supercooled water were found.

In a summary of Project Stormfury (Sheets 1981) the existing studies of natural storm variability, exploratory seeding experiments, and numerical and theoretical simulations indicate that the implementation of the Stormfury hypothesis could result in reductions of 10–15 per cent in the maximum wind speed, with associated damage reductions of 20–60 per cent. No significant changes in storm motion nor storm averaged rainfall at any specific location should be expected.

Currently, further seeding experiments are not being performed, although detailed observational sampling of tropical storms is continuing as part of the Stormfury project. It is likely that further seeding experiments will be conducted only after a major catastrophe results from a landfalling hurricane in the United States.

Appendix A

ATLANTIC TROPICAL CYCLONE TRACKS, 1871–1989

Source: Neumann, Jarvinen and Pike, 1987

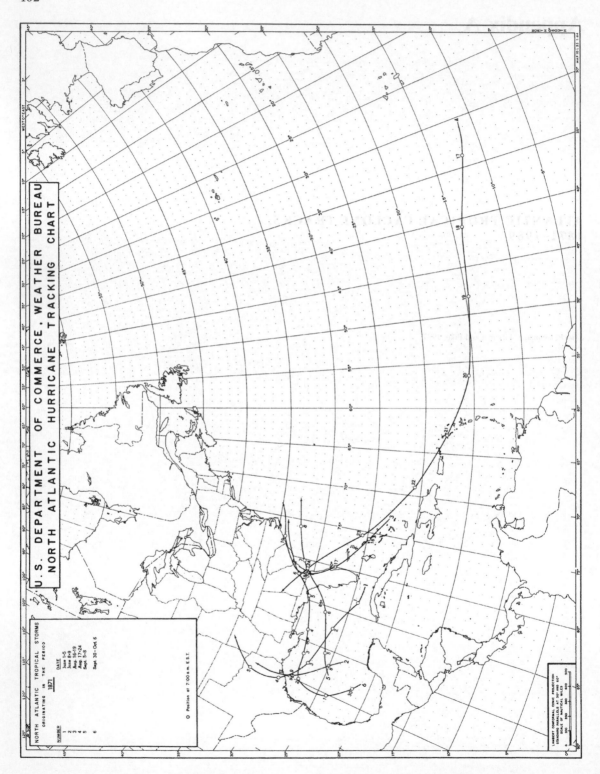

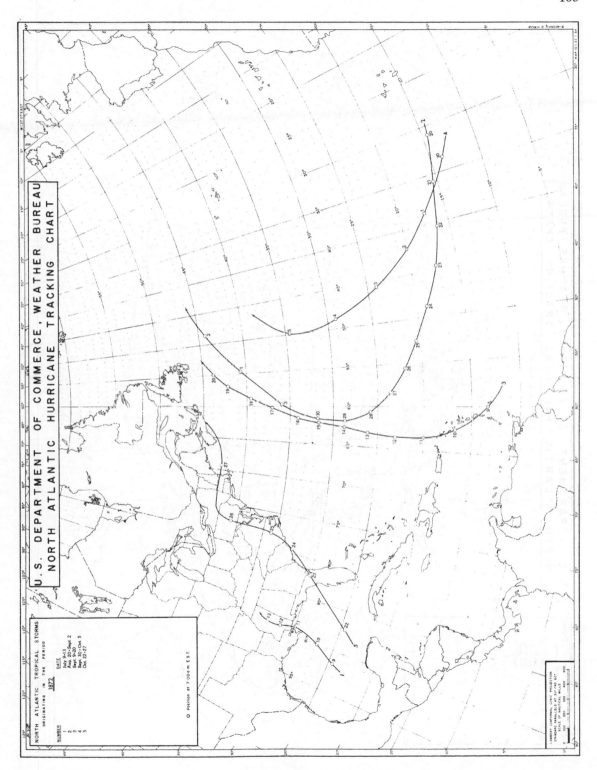

NORTH ATLANTIC TROPICAL STORMS
ORIGINATING IN THE PERIOD
1872

U. S. DEPARTMENT OF COMMERCE, WEATHER BUREAU
NORTH ATLANTIC HURRICANE TRACKING CHART

NUMBER	DATE
1	July 9-13
2	Aug. 20-Sept. 2
3	Sept. 9-20
4	Sept. 30-Oct. 5
5	Oct. 22-27

O Position at 7:00 a.m. EST

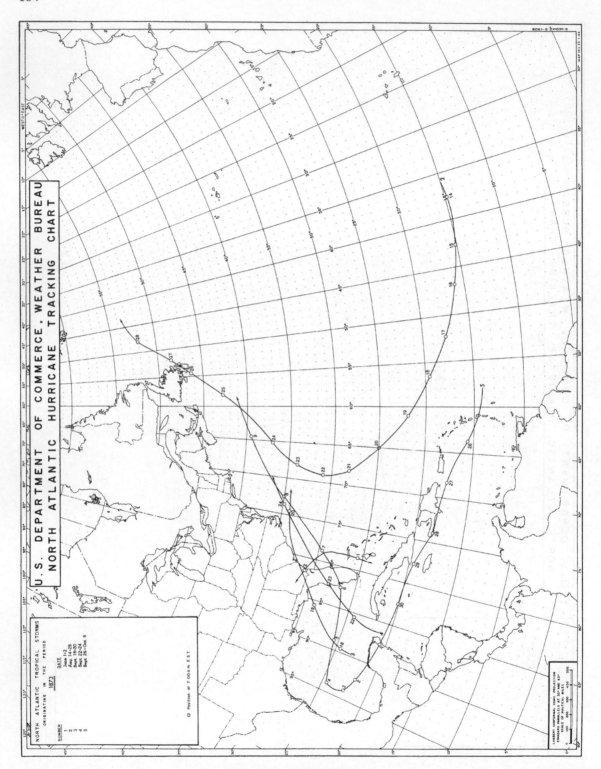

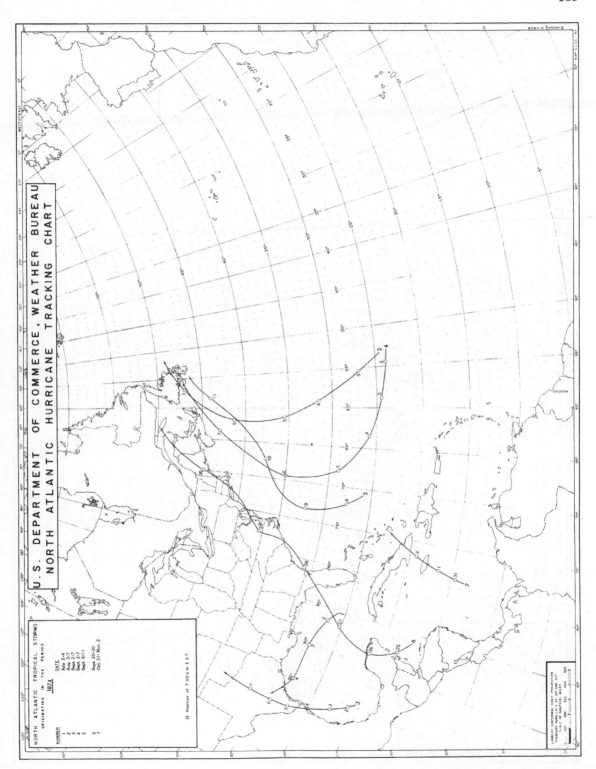

U.S. DEPARTMENT OF COMMERCE, WEATHER BUREAU
NORTH ATLANTIC HURRICANE TRACKING CHART

NORTH ATLANTIC TROPICAL STORMS
ORIGINATING IN THE PERIOD
1874

NUMBER	DATE
1	July 2-4
2	Aug. 3-7
3	Sept. 2-7
4	Sept. 2-7
5	Sept. 8-11
6	Sept. 25-30
7	Oct. 31-Nov. 2

O Position at 7:00 am E.S.T.

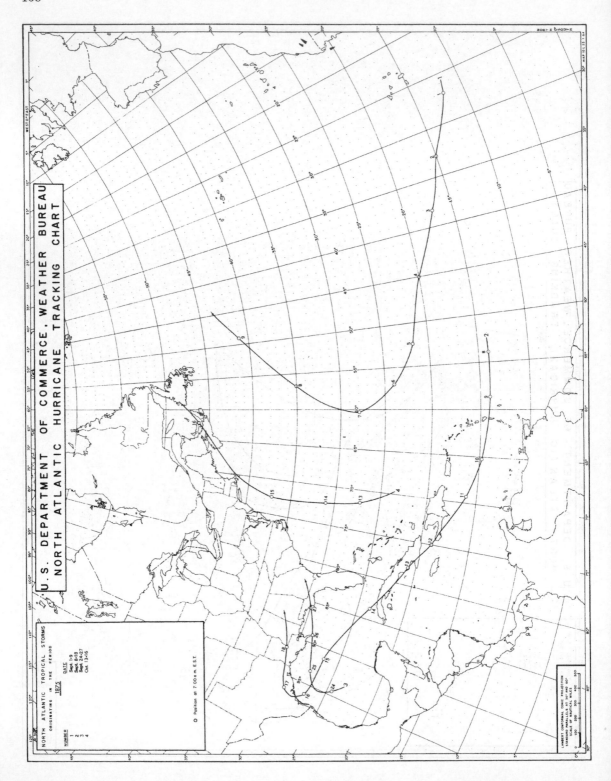

U.S. DEPARTMENT OF COMMERCE, WEATHER BUREAU
NORTH ATLANTIC HURRICANE TRACKING CHART

NORTH ATLANTIC TROPICAL STORMS
ORIGINATING IN THE PERIOD
1875

NUMBER   DATE
1   Sept. 1-9
2   Sept. 8-18
3   Sept. 24-27
4   Oct. 13-16

○ Position at 7:00 a.m. E.S.T.

LAMBERT CONFORMAL CONIC PROJECTION
STANDARD PARALLELS AT 30° AND 60°
SCALE OF NAUTICAL MILES

107

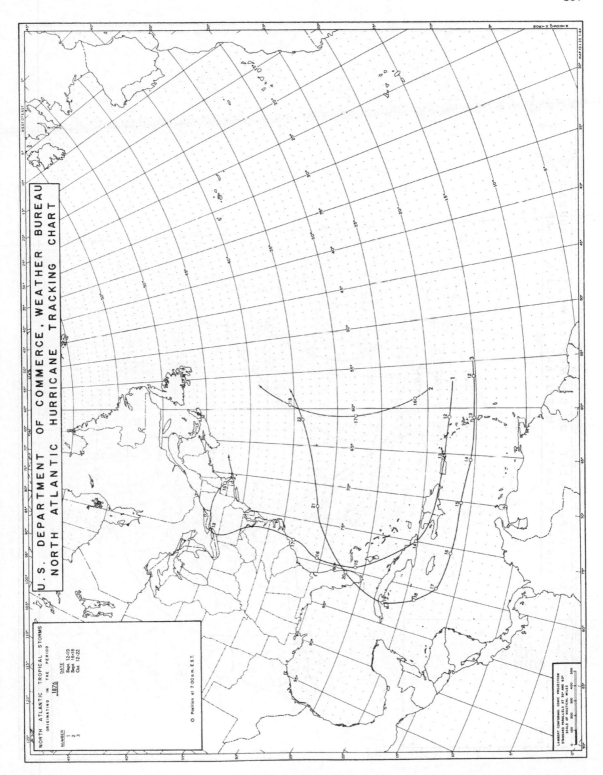

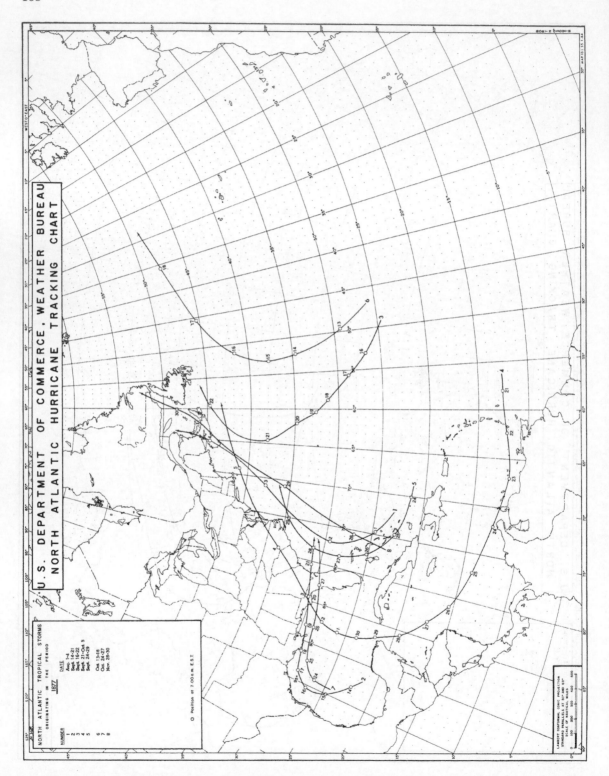

U.S. DEPARTMENT OF COMMERCE, WEATHER BUREAU
NORTH ATLANTIC HURRICANE TRACKING CHART

NORTH ATLANTIC TROPICAL STORMS
ORIGINATING IN THE PERIOD
1877

NUMBER	DATE
1	Aug. 1-4
2	Sept. 14-21
3	Sept. 16-22
4	Sept. 21-Oct. 5
5	Sept. 24-29
6	Oct. 13-18
7	Oct. 24-27
8	Nov. 28-30

O Position at 7:00 a.m. E.S.T.

LAMBERT CONFORMAL CONIC PROJECTION
STANDARD PARALLELS AT 30° AND 60°
SCALE OF NAUTICAL MILES

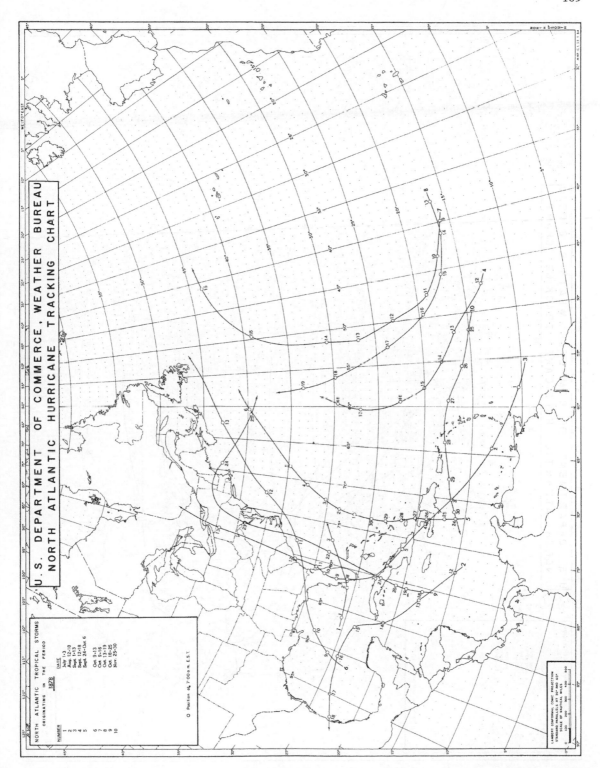

U.S. DEPARTMENT OF COMMERCE, WEATHER BUREAU
NORTH ATLANTIC HURRICANE TRACKING CHART

NORTH ATLANTIC TROPICAL STORMS
ORIGINATING IN THE PERIOD
1928

NUMBER	DATE
1	July 1-3
2	Aug. 12-18
3	Sept. 1-13
4	Sept. 12-18
5	Sept. 24-Oct. 6
6	Oct. 9-13
7	Oct. 9-16
8	Oct. 13-19
9	Oct. 18-25
10	Nov. 25-30

O Position at 7:00 a.m. E.S.T.

LAMBERT CONFORMAL CONIC PROJECTION
STANDARD PARALLELS AT 30° AND 65°
SCALE OF NAUTICAL MILES

110

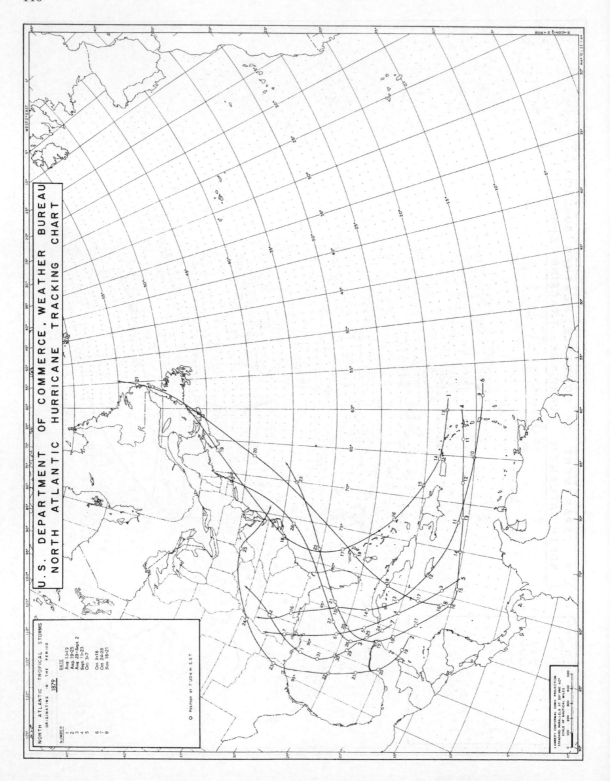

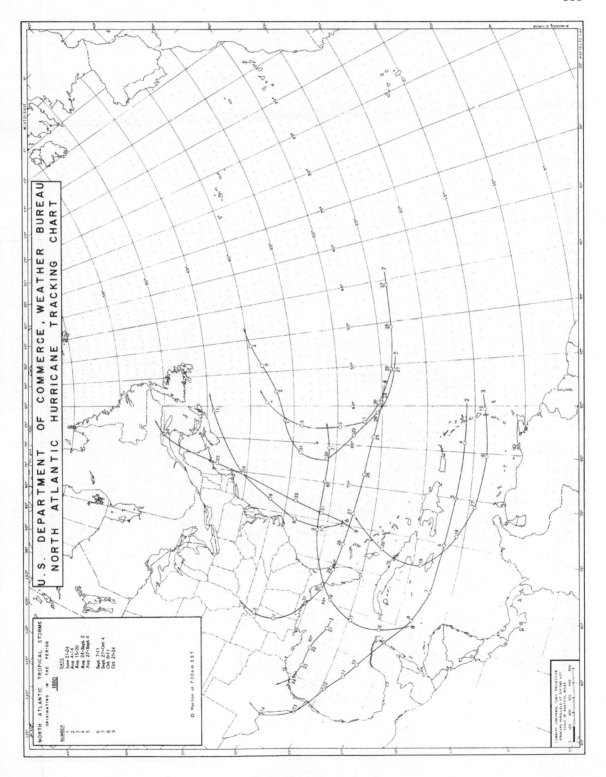

U. S. DEPARTMENT OF COMMERCE, WEATHER BUREAU
NORTH ATLANTIC HURRICANE TRACKING CHART

NORTH ATLANTIC TROPICAL STORMS
ORIGINATING IN THE PERIOD
1880

NUMBER	DATE
1	June 21-24
2	Aug. 4-14
3	Aug. 15-20
4	Aug. 24-Sept. 2
5	Aug. 27-Sept. 4
6	Sept. 7-11
7	Sept. 27-Oct. 4
8	Oct. 6-11
9	Oct. 21-24

O Position at 7:00 a.m. EST

112

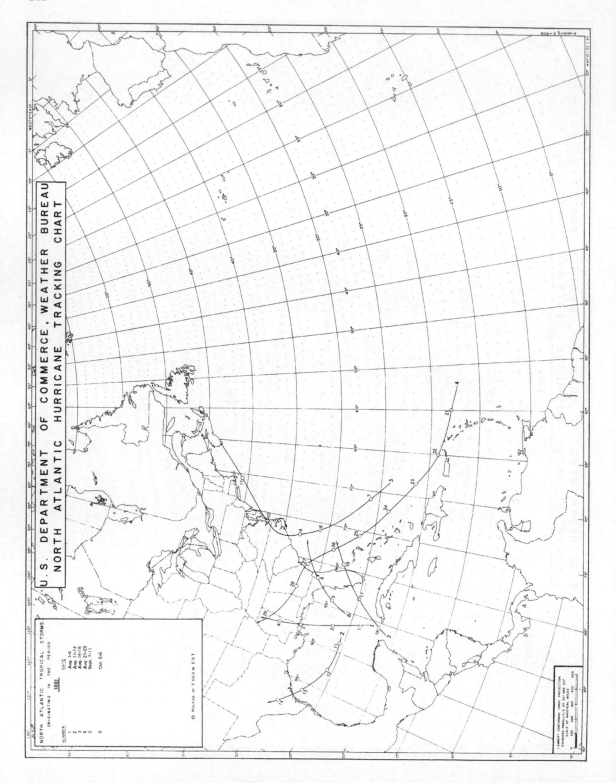

113

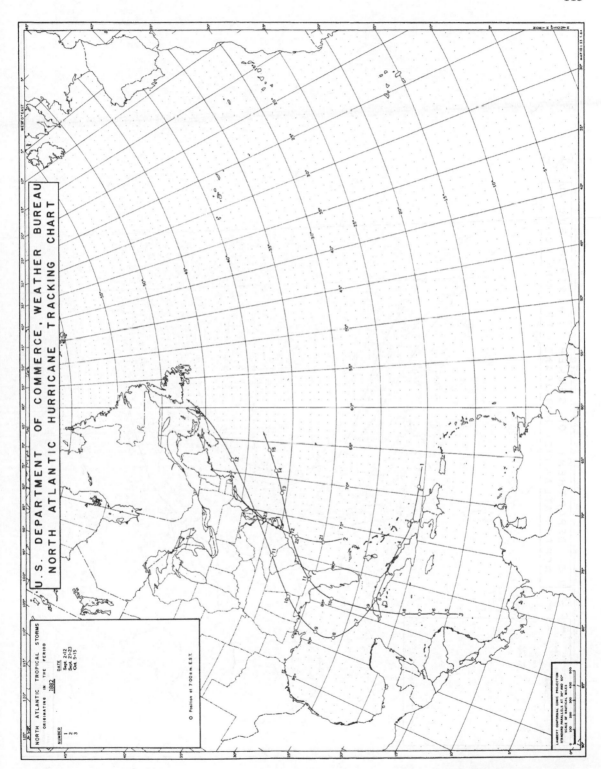

114

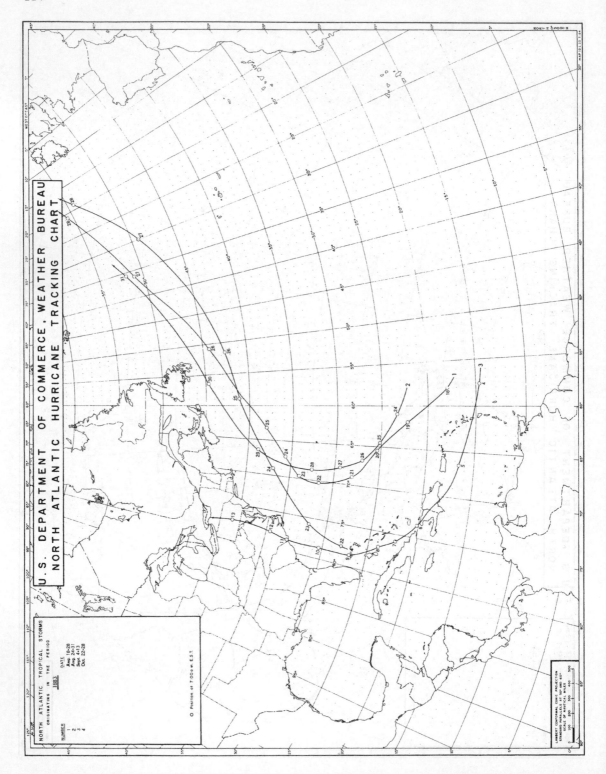

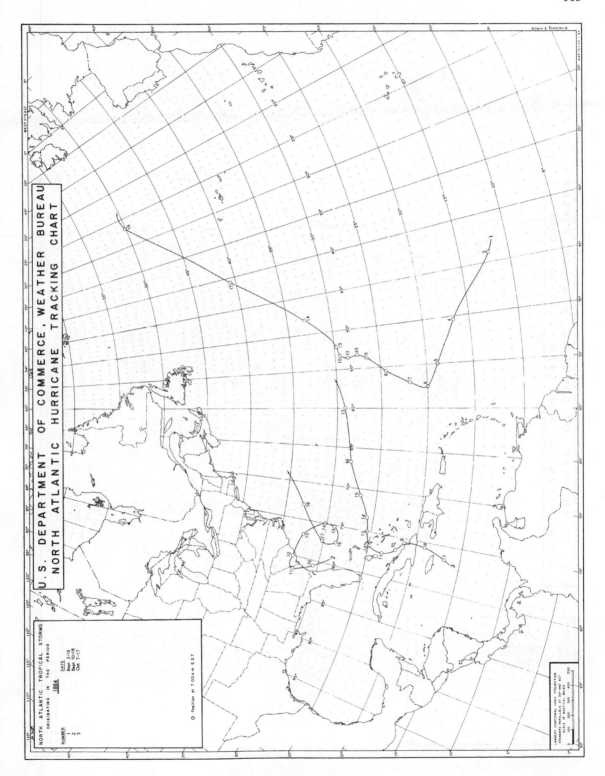

116

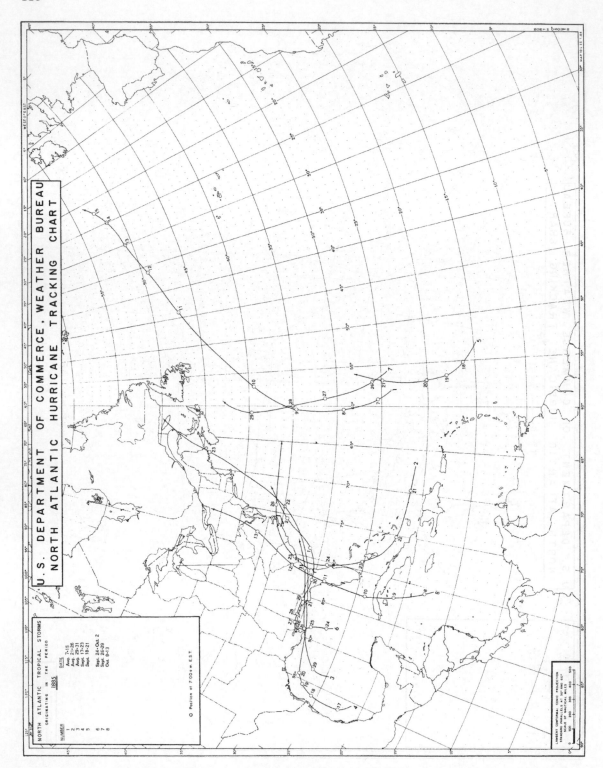

U. S. DEPARTMENT OF COMMERCE, WEATHER BUREAU
NORTH ATLANTIC HURRICANE TRACKING CHART

NORTH ATLANTIC TROPICAL STORMS
ORIGINATING IN THE PERIOD
1885

NUMBER	DATE
1	Aug. 7-15
2	Aug. 21-26
3	Aug. 29-31
4	Sept. 17-23
5	Sept. 18-21
6	Sept. 24-Oct. 2
7	Sept. 26-29
8	Oct. 9-13

○ Position at 7:00 a.m. EST

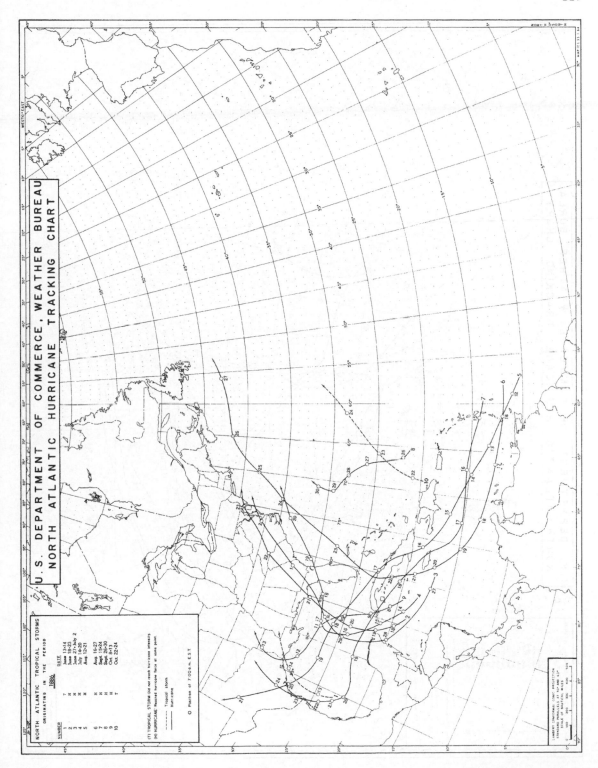

118

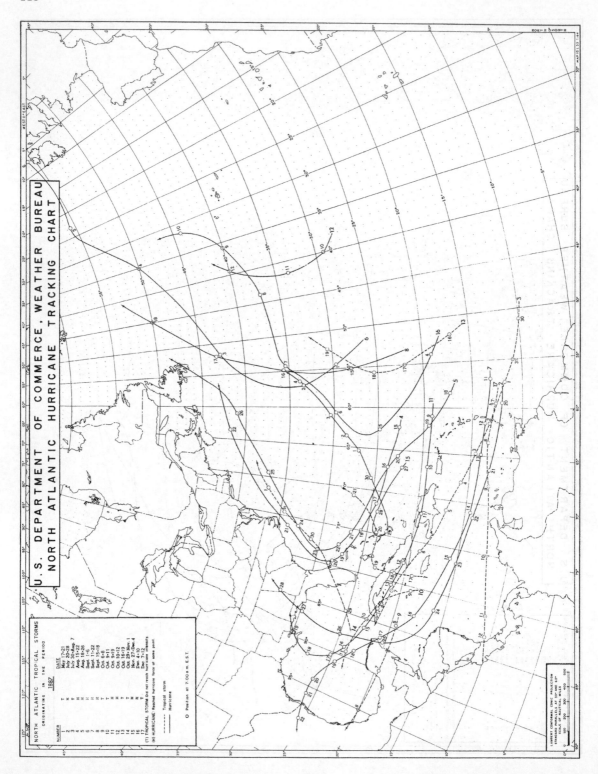

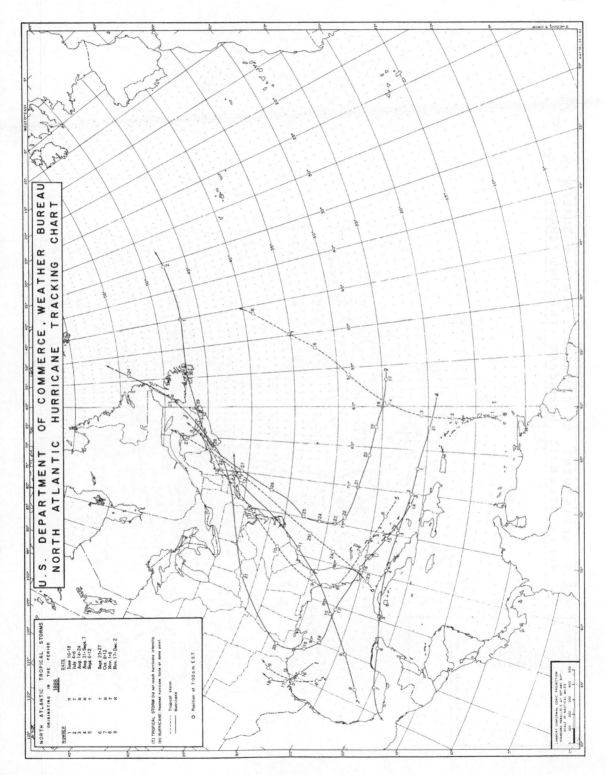

NORTH ATLANTIC TROPICAL STORMS
ORIGINATING IN THE PERIOD
1888

NUMBER		DATE
1	H	June 16-18
2	T	July 4-6
3	H	Aug. 14-24
4	H	Aug. 31-Sept. 7
5	T	Sept. 6-12
6	H	Sept. 23-27
7	H	Oct. 9-12
8	T	Nov. 1-8
9	H	Nov. 17-Dec. 2

(T) TROPICAL STORM Did not reach hurricane intensity.

(H) HURRICANE Reached hurricane force at some point.

—— Tropical storm
- - - - Hurricane

○ Position at 7:00 a.m. E.S.T.

U. S. DEPARTMENT OF COMMERCE, WEATHER BUREAU
NORTH ATLANTIC HURRICANE TRACKING CHART

LAMBERT CONFORMAL CONIC PROJECTION
STANDARD PARALLELS AT 9° AND 60°
SCALE OF NAUTICAL MILES

120

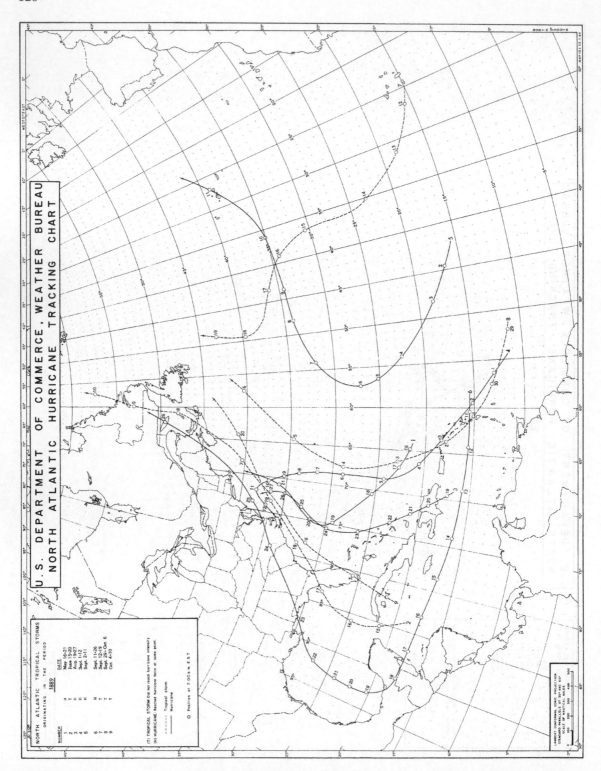

U.S. DEPARTMENT OF COMMERCE, WEATHER BUREAU
NORTH ATLANTIC HURRICANE TRACKING CHART

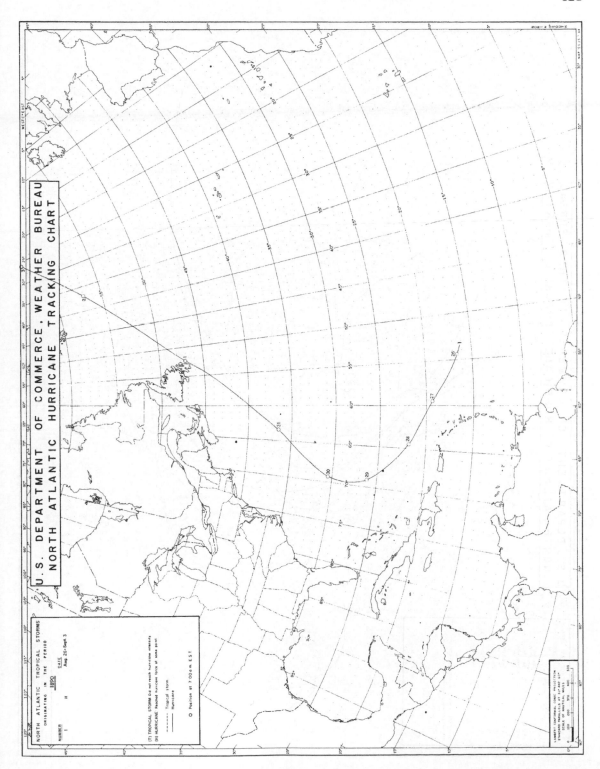

U.S. DEPARTMENT OF COMMERCE, WEATHER BUREAU
NORTH ATLANTIC HURRICANE TRACKING CHART

NORTH ATLANTIC TROPICAL STORMS
ORIGINATING IN THE PERIOD
1890

NUMBER   DATE
1   Aug. 26-Sept. 3

(T) TROPICAL STORM Did not reach hurricane intensity
(H) HURRICANE Reached hurricane force at some point
------- Tropical storm
———— Hurricane
O Position at 7:00 a.m. EST

122

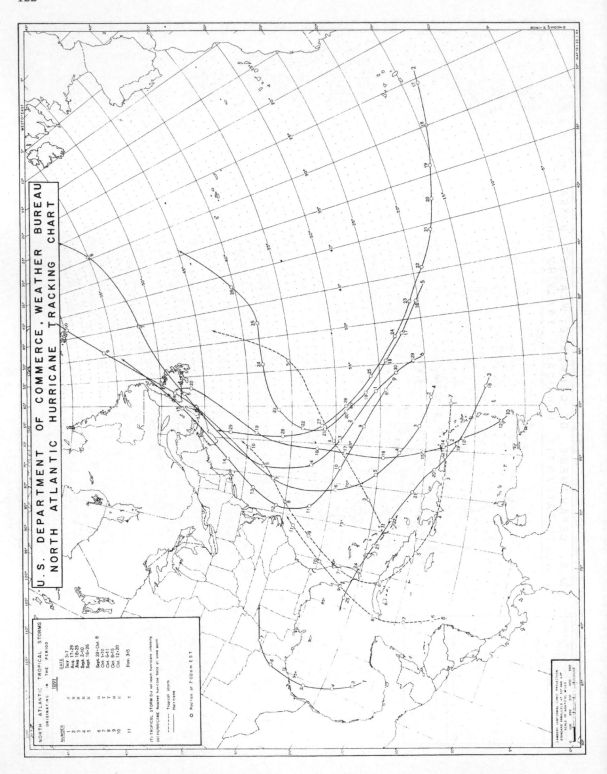

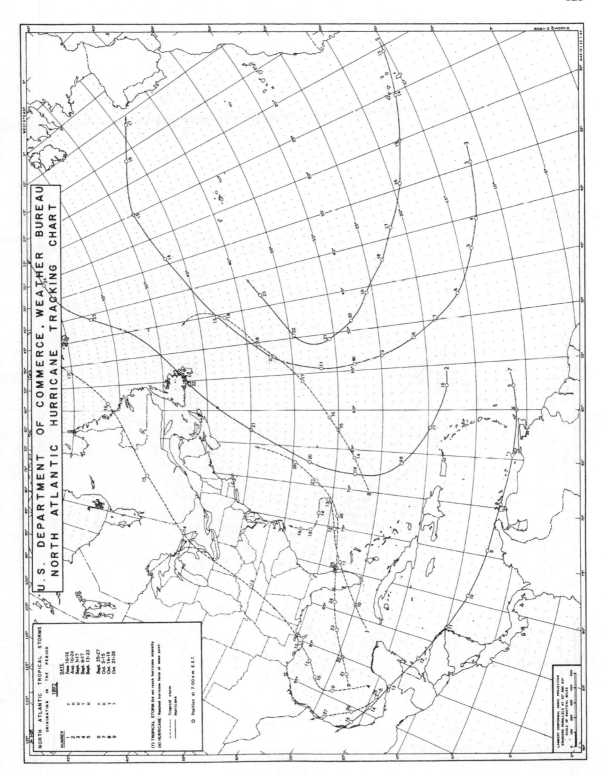

U.S. DEPARTMENT OF COMMERCE, WEATHER BUREAU
NORTH ATLANTIC HURRICANE TRACKING CHART

NORTH ATLANTIC TROPICAL STORMS
ORIGINATING IN THE PERIOD
1892

NUMBER		DATE
1	T	**June** 10-16
2	H	**Aug.** 16-24
3	H	**Sept.** 3-17
4	T	**Sept.** 9-17
5	H	**Sept.** 13-23
6	T	**Sept.** 25-27
7	H	**Oct.** 5-15
8	T	**Oct.** 14-19
9	T	**Oct.** 21-28

(T) TROPICAL STORM Did not reach hurricane intensity.
(H) HURRICANE Reached hurricane force at some point.

- - - - - Tropical storm
———— Hurricane

O Position at 7:00 a.m. E.S.T.

LAMBERT CONFORMAL CONIC PROJECTION
STANDARD PARALLELS AT 20° AND 60°
SCALE OF NAUTICAL MILES

124

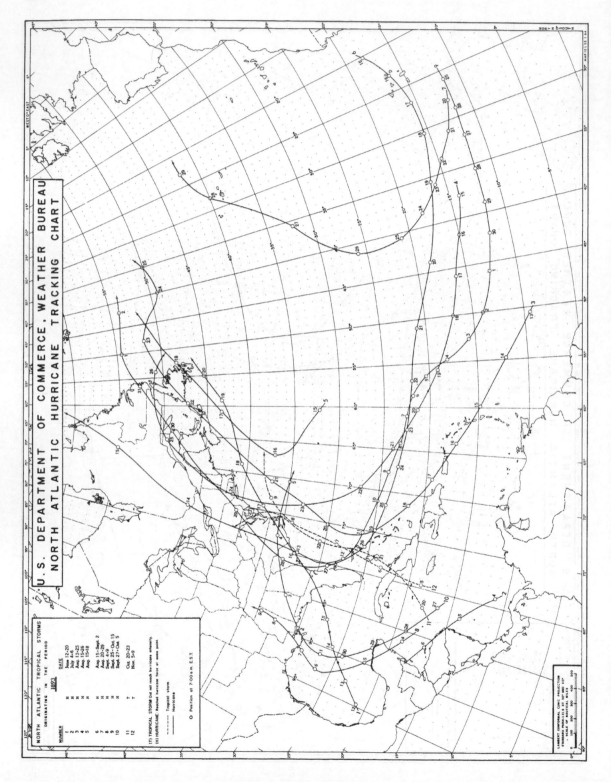

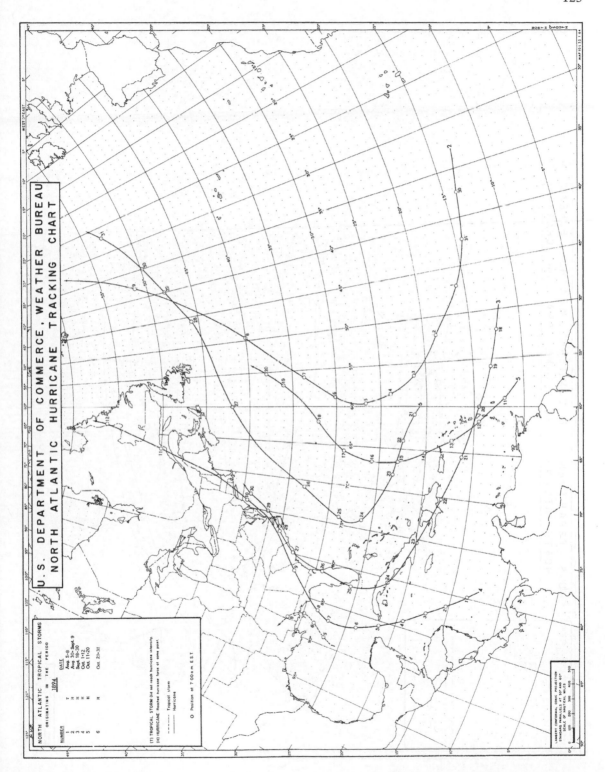

U.S. DEPARTMENT OF COMMERCE, WEATHER BUREAU
NORTH ATLANTIC HURRICANE TRACKING CHART

NORTH ATLANTIC TROPICAL STORMS
ORIGINATING IN THE PERIOD
1894

126

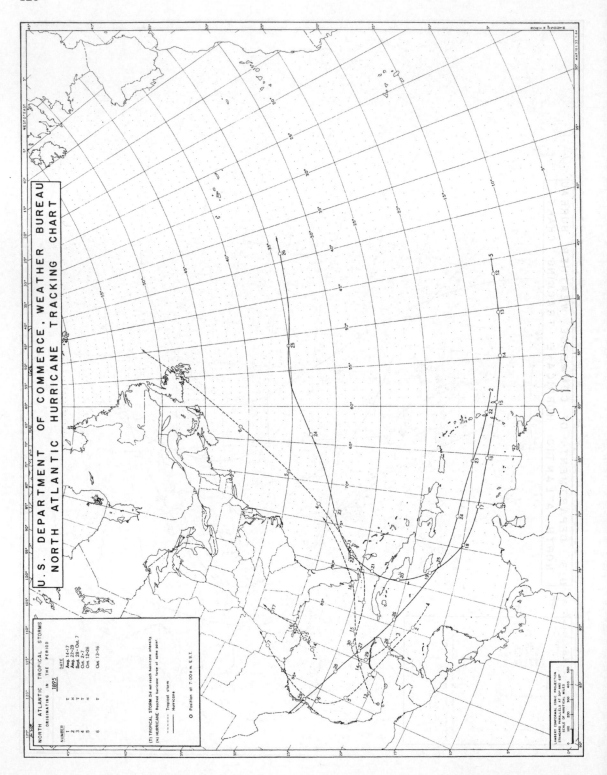

U.S. DEPARTMENT OF COMMERCE, WEATHER BUREAU
NORTH ATLANTIC HURRICANE TRACKING CHART

NORTH ATLANTIC TROPICAL STORMS
ORIGINATING IN THE PERIOD
1895

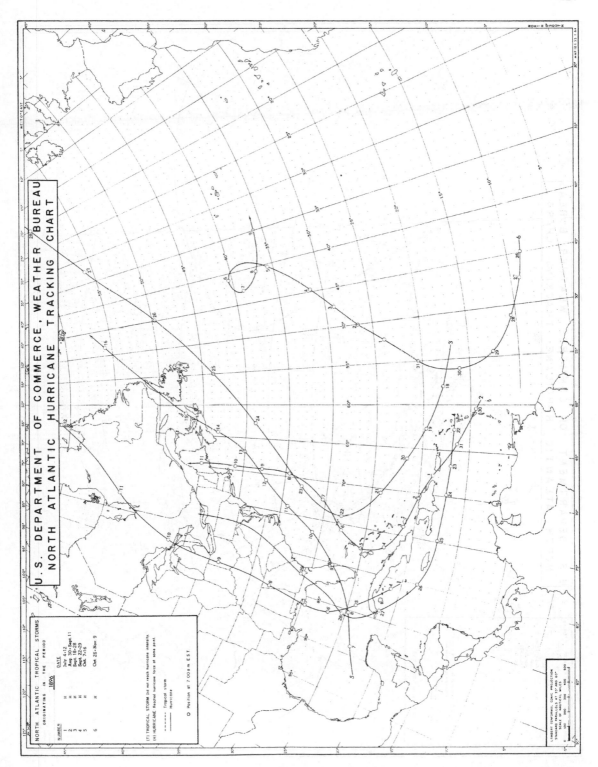

U.S. DEPARTMENT OF COMMERCE, WEATHER BUREAU
NORTH ATLANTIC HURRICANE TRACKING CHART

NORTH ATLANTIC TROPICAL STORMS
ORIGINATING IN THE PERIOD
1892

NUMBER		DATE
1	H	July 4-12
2	H	Aug. 30-Sept. 11
3	H	Sept. 18-28
4	H	Sept. 22-29
5	H	Oct. 7-16
6	H	Oct. 26-Nov. 9

(T) TROPICAL STORM Did not reach hurricane intensity.

(H) HURRICANE Reached hurricane force at some point.

‑ ‑ ‑ ‑ ‑ Tropical storm

———— Hurricane

O Position at 7:00 a.m E.S.T.

128

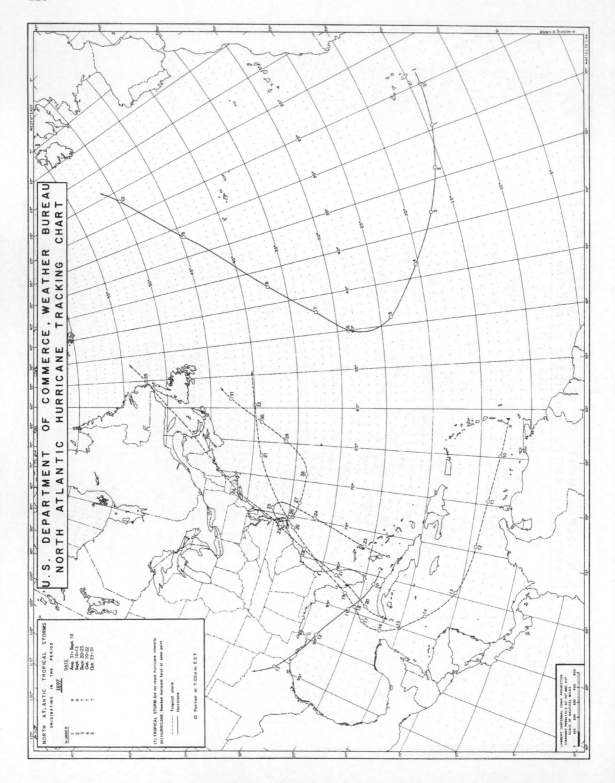

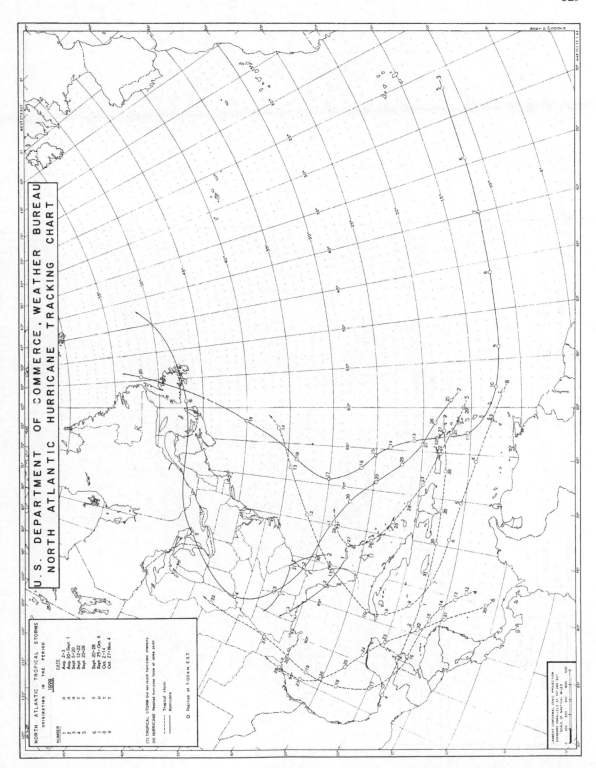

U.S. DEPARTMENT OF COMMERCE, WEATHER BUREAU
NORTH ATLANTIC HURRICANE TRACKING CHART

NORTH ATLANTIC TROPICAL STORMS
ORIGINATING IN THE PERIOD
1898

NUMBER		DATE
1	H	Aug. 2-3
2	H	Aug. 30-Sept. 1
3	H T	Sept. 5-20
4	T	Sept. 12-22
5	T	Sept. 20-28
6	T	Sept. 20-28
7	H T	Sept. 25-Oct. 6
8	H T	Oct. 2-14
9	T	Oct. 27-Nov. 4

(T) TROPICAL STORM Did not reach hurricane strength
(H) HURRICANE Reached hurricane force at some point

——— Tropical Storm
------- Hurricane

O Position at 7:00 a.m. E.S.T.

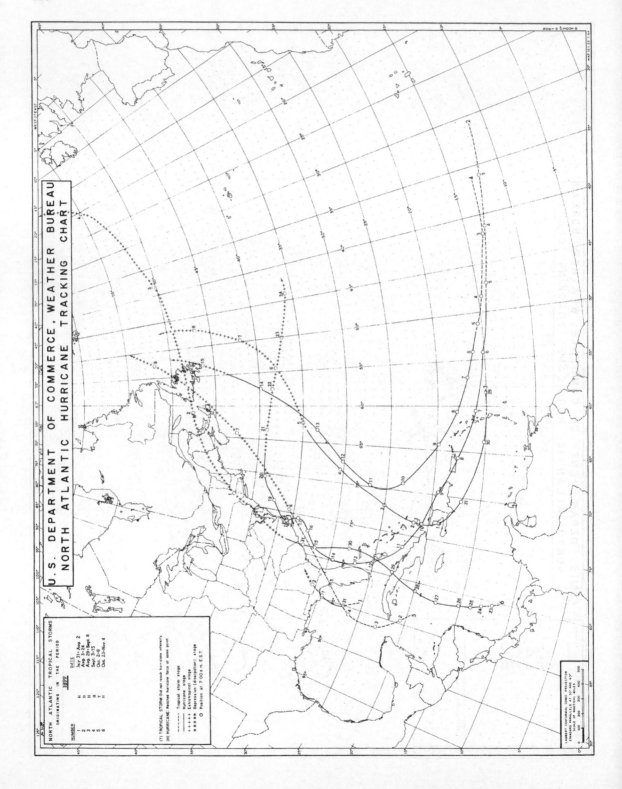

U.S. DEPARTMENT OF COMMERCE, WEATHER BUREAU
NORTH ATLANTIC HURRICANE TRACKING CHART

NORTH ATLANTIC TROPICAL STORMS
ORIGINATING IN THE PERIOD

1892

NUMBER	DATE
1	July 31–Aug. 2
2	Aug. 3–24
3	Aug. 29–Sept. 8
4	Sept. 3–15
5	Oct. 2–8
6	Oct. 23–Nov. 4

(T) TROPICAL STORM: Did not reach hurricane intensity
(H) HURRICANE: Reached hurricane force at some point

Tropical storm stage
Hurricane stage
Extratropical stage
Depression (dissipation) stage
Position at 7:00 a.m. EST

LAMBERT CONFORMAL CONIC PROJECTION
STANDARD PARALLELS AT 30° AND 60°
SCALE OF NAUTICAL MILES

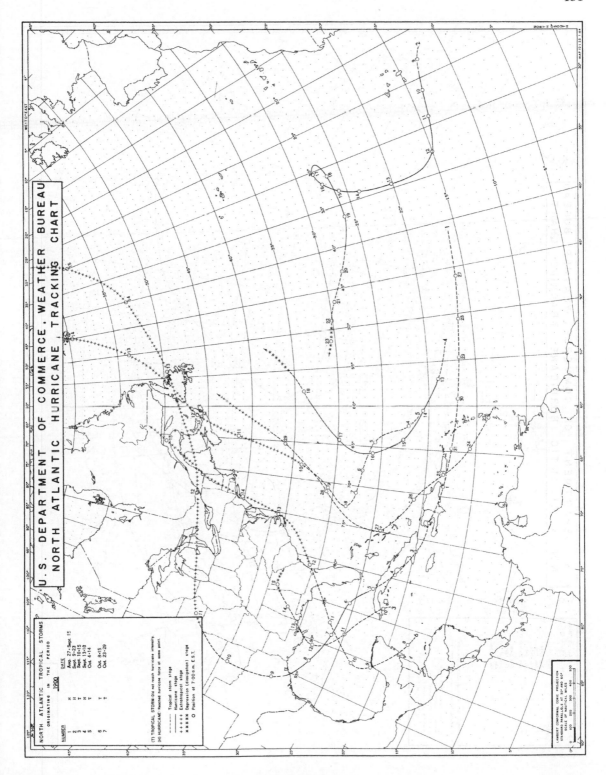

132

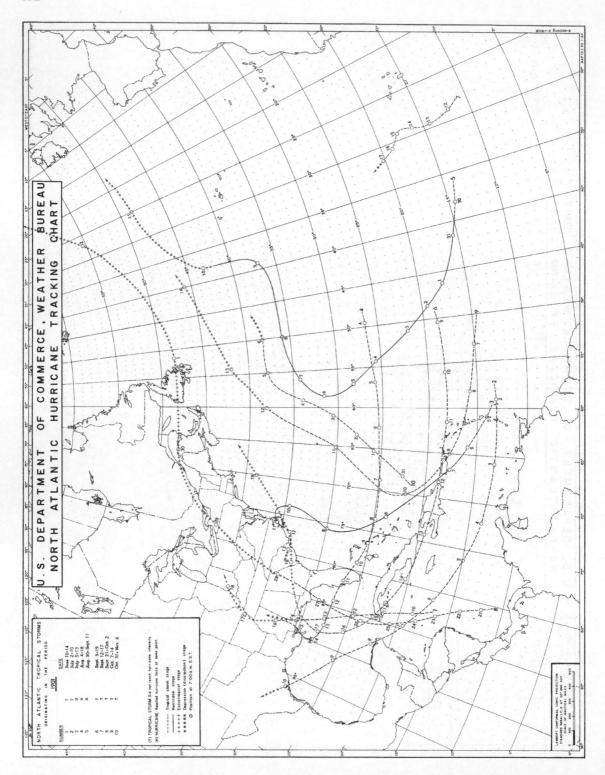

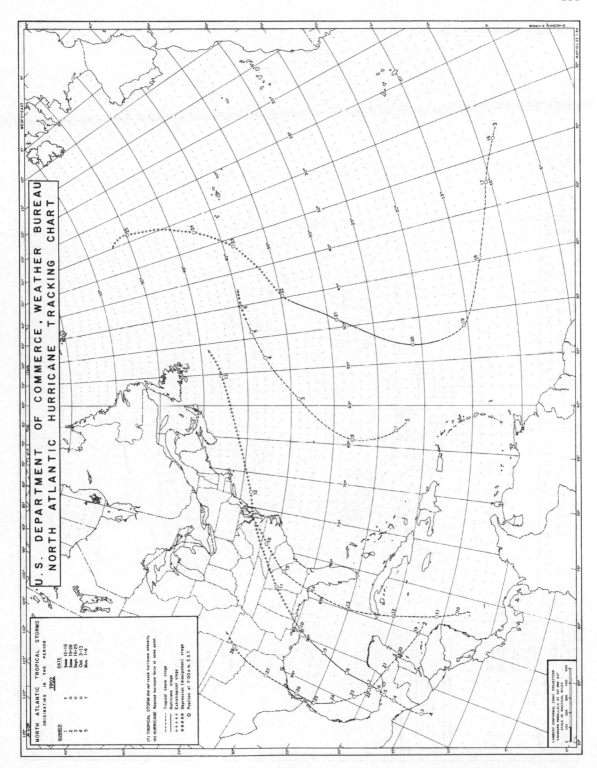

U.S. DEPARTMENT OF COMMERCE, WEATHER BUREAU
NORTH ATLANTIC HURRICANE TRACKING CHART

NORTH ATLANTIC TROPICAL STORMS
ORIGINATING IN THE PERIOD
1902

NUMBER		DATE
1	T	June 10-16
2	H	June 19-28
3	H	Sept. 16-25
4	H	Oct. 3-13
5	T	Nov. 1-6

(T) TROPICAL STORM: Did not reach hurricane intensity.
(H) HURRICANE: Reached hurricane force at some point.

Tropical storm stage
Hurricane stage
+++++ Extratropical stage
xxxxx Depression (dissipation) stage
○ Position at 7:00 a.m. E.S.T.

134

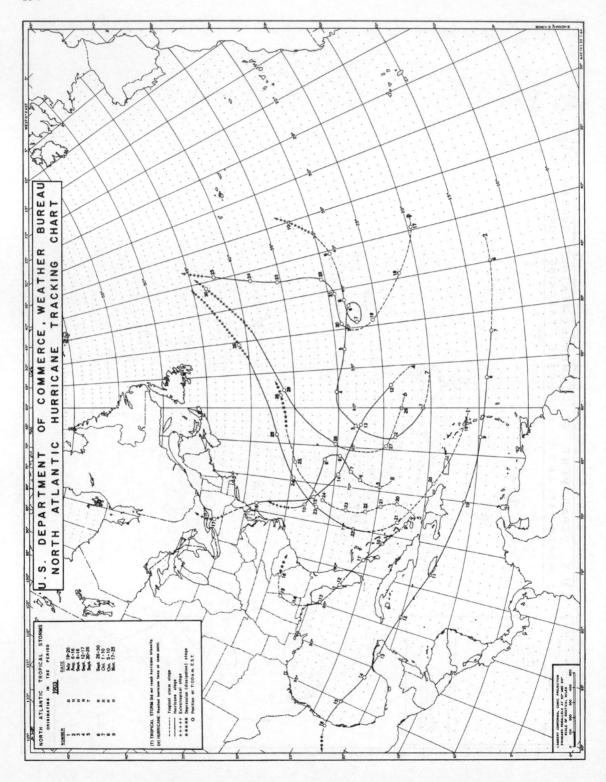

U.S. DEPARTMENT OF COMMERCE, WEATHER BUREAU
NORTH ATLANTIC HURRICANE TRACKING CHART

NORTH ATLANTIC TROPICAL STORMS
ORIGINATING IN THE PERIOD
1903

NUMBER	DATE
1 H	July 19-26
2 H	Aug. 6-16
3 H	Aug. 9-16
4 T	Sept. 12-17
5 H	Sept. 20-25
6 H	Sept. 26-30
7 H	Oct. 1-10
8 H	Oct. 5-10
9 H	Nov. 17-25

(T) TROPICAL STORM: Did not reach hurricane intensity.
(H) HURRICANE: Reached hurricane force of same point.

+++++ Tropical storm stage
+++++ Hurricane stage
+++++ Extratropical stage
××××× Depression (dissipation) stage
O Position at 7:00 a.m. E.S.T.

LAMBERT CONFORMAL CONIC PROJECTION
STANDARD PARALLELS AT 9° AND 49°
SCALE OF NAUTICAL MILES

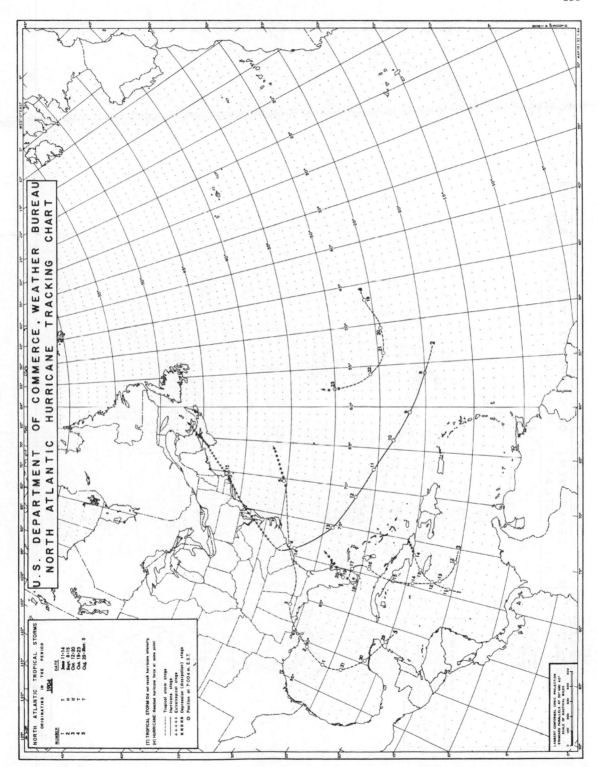

U.S. DEPARTMENT OF COMMERCE, WEATHER BUREAU
NORTH ATLANTIC HURRICANE TRACKING CHART

136

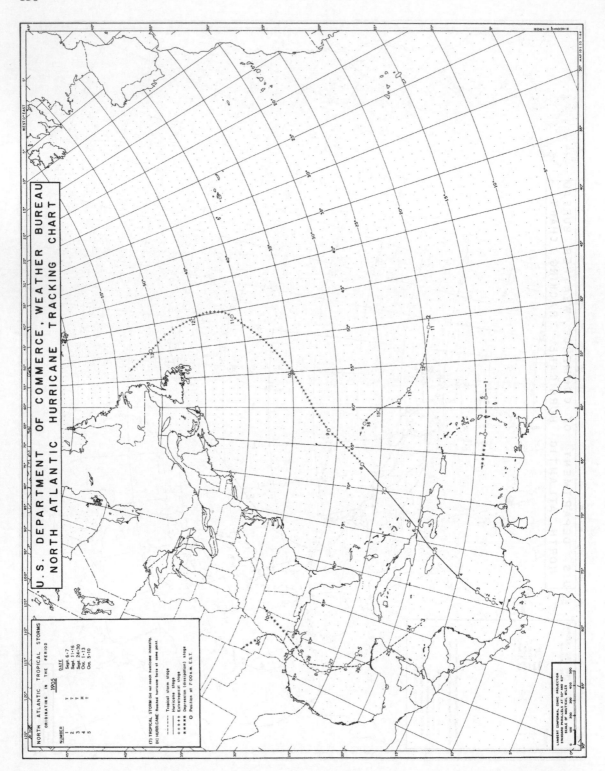

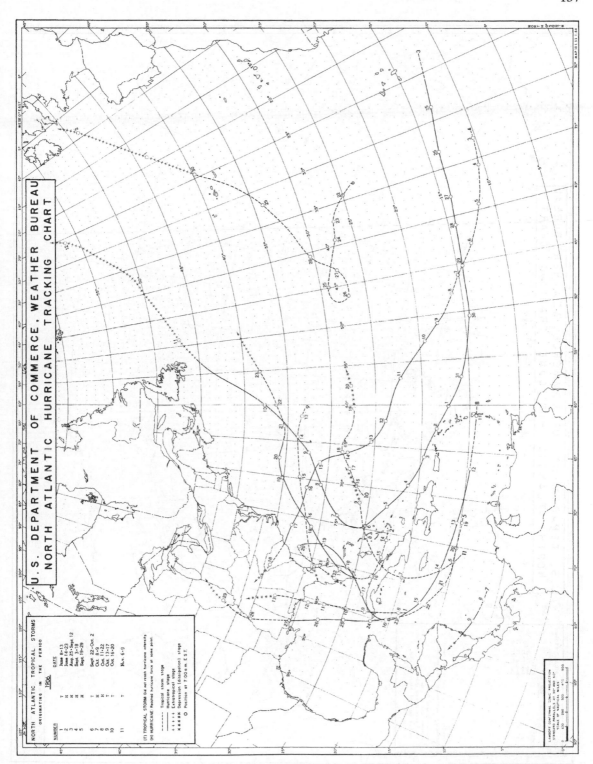

138

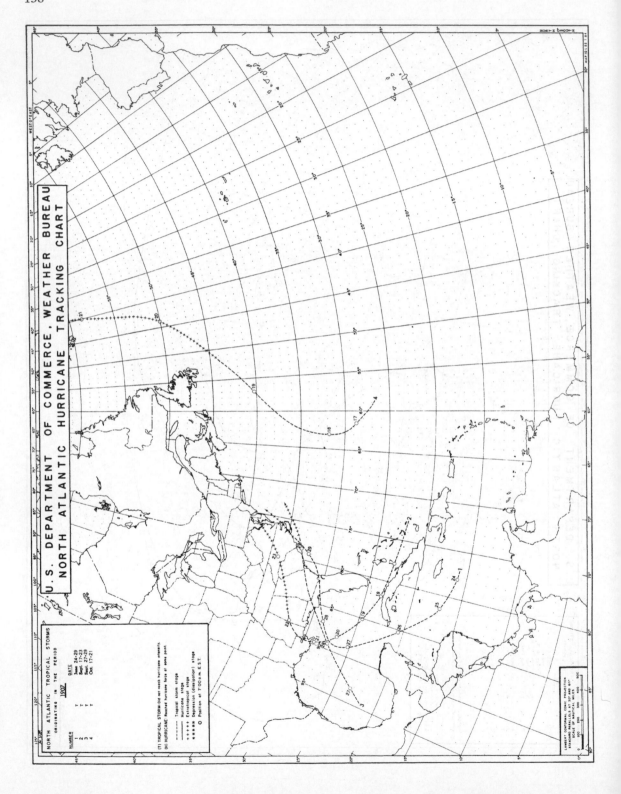

U.S. DEPARTMENT OF COMMERCE, WEATHER BUREAU
NORTH ATLANTIC HURRICANE TRACKING CHART

139

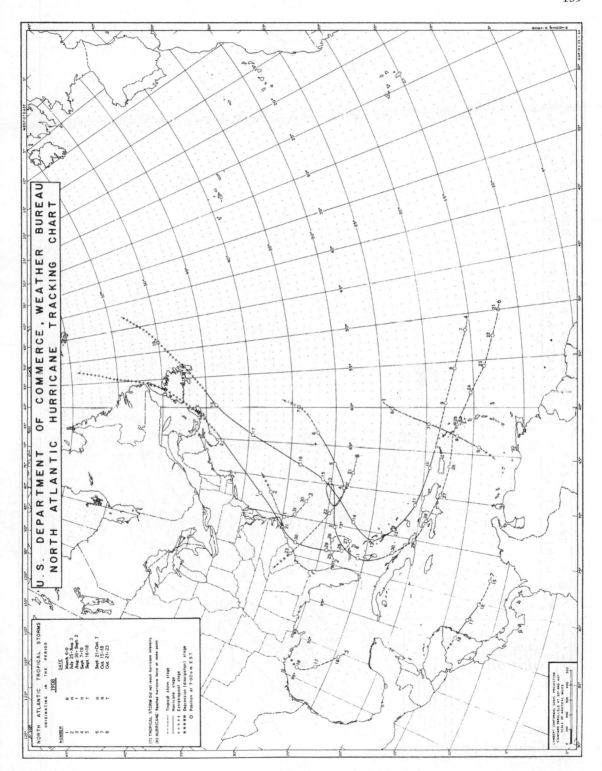

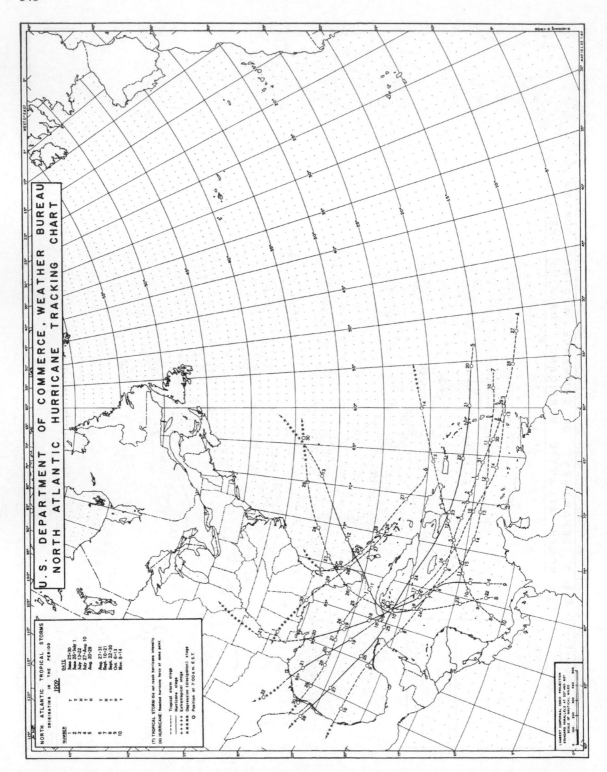

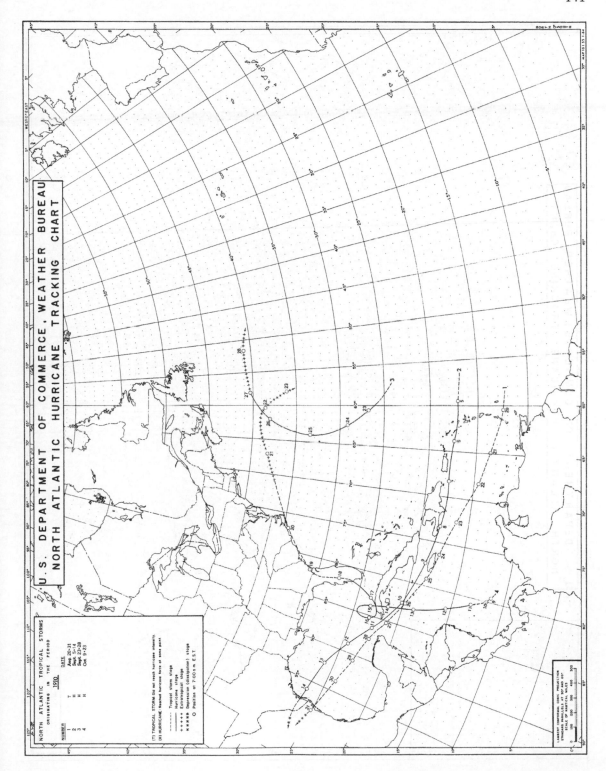

U. S. DEPARTMENT OF COMMERCE, WEATHER BUREAU
NORTH ATLANTIC HURRICANE TRACKING CHART

NORTH ATLANTIC TROPICAL STORMS
ORIGINATING IN THE PERIOD
1910

NUMBER	DATE
1	T	Aug. 20-31
2	H	Sept. 5-14
3	H	Sept. 23-28
4	H	Oct. 9-23

(T) TROPICAL STORM. Did not reach hurricane intensity.
(H) HURRICANE. Reached hurricane force at some point.

- - - - - Tropical storm stage
━━━━━ Hurricane stage
+ + + + + Extratropical stage
н н н н Depression (dissipation) stage
○ Position at 7:00 a.m. E.S.T.

142

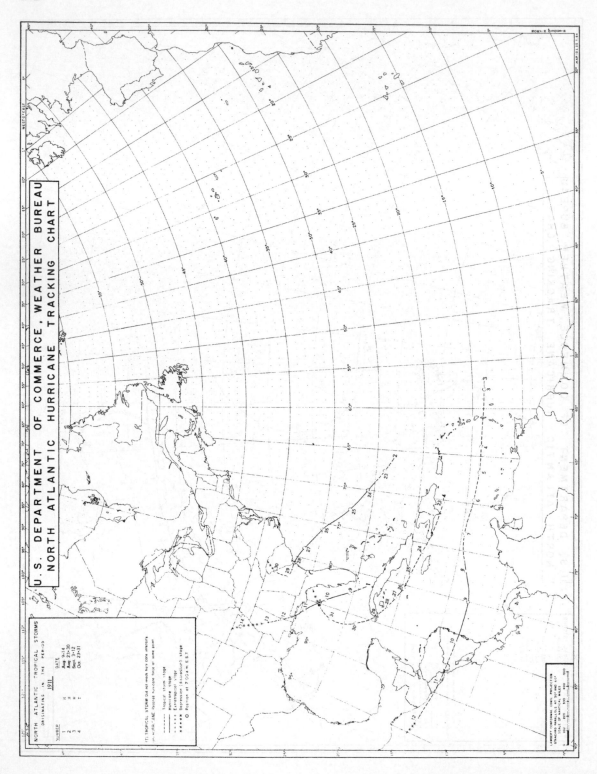

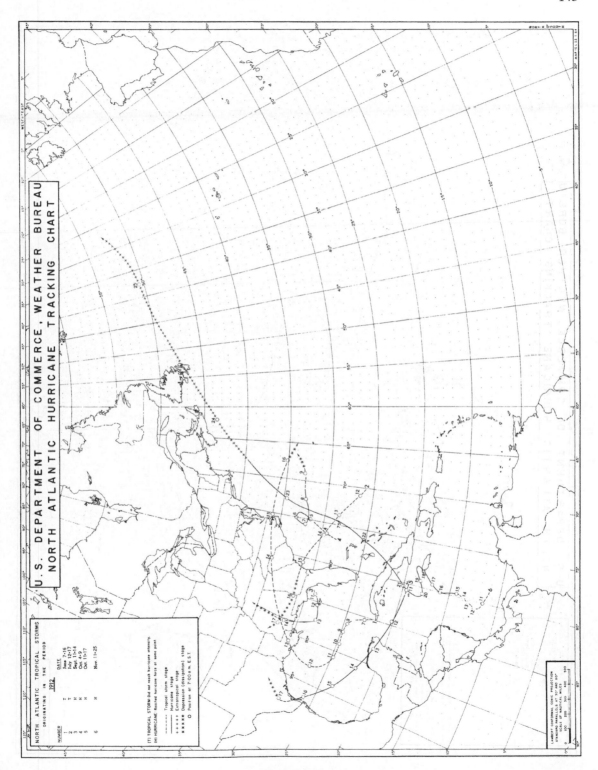

U.S. DEPARTMENT OF COMMERCE, WEATHER BUREAU
NORTH ATLANTIC HURRICANE TRACKING CHART

NORTH ATLANTIC TROPICAL STORMS
ORIGINATING IN THE PERIOD
1912

NUMBER		DATE
1	T	June 7-16
2	T	July 12-17
3	H	Sept. 11-14
4	H	Oct. 4-9
5	H	Oct. 11-17
6	H	Nov. 11-25

(T) TROPICAL STORM: Did not reach hurricane intensity.

(H) HURRICANE: Reached hurricane force at some point

– – – – – Tropical storm stage
━━━━━ Hurricane stage
+ + + + + Extratropical stage
✕ ✕ ✕ ✕ ✕ Depression (dissipation) stage
○ Position at 7:00 a.m. E.S.T.

LAMBERT CONFORMAL CONIC PROJECTION
STANDARD PARALLELS AT 20° AND 60°
SCALE OF NAUTICAL MILES

144

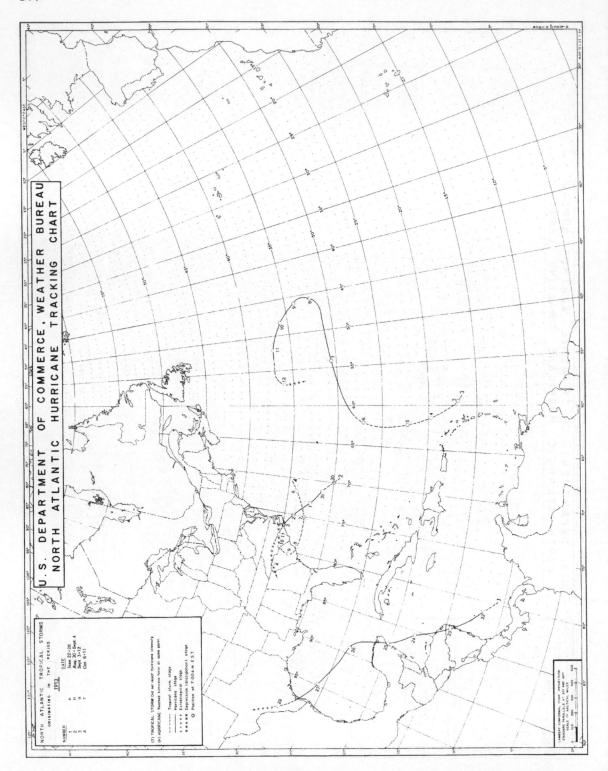

NORTH ATLANTIC TROPICAL STORMS
ORIGINATING IN THE PERIOD
1913

NUMBER		DATE
1	H	June 22-28
2	H	Aug. 30-Sept. 4
3	H	Sept. 3-17
4	T	Oct. 6-11

(T) TROPICAL STORM Did not reach hurricane intensity.
(H) HURRICANE Reached hurricane force at some point.

——— Tropical storm stage
———— Hurricane stage
+ + + + Extratropical stage
* * * * Depression (dissipation) stage
O Position at 7:00 a.m. E.S.T.

U.S. DEPARTMENT OF COMMERCE, WEATHER BUREAU
NORTH ATLANTIC HURRICANE TRACKING CHART

LAMBERT CONFORMAL CONIC PROJECTION
STANDARD PARALLELS AT 30° AND 60°
SCALE OF NAUTICAL MILES

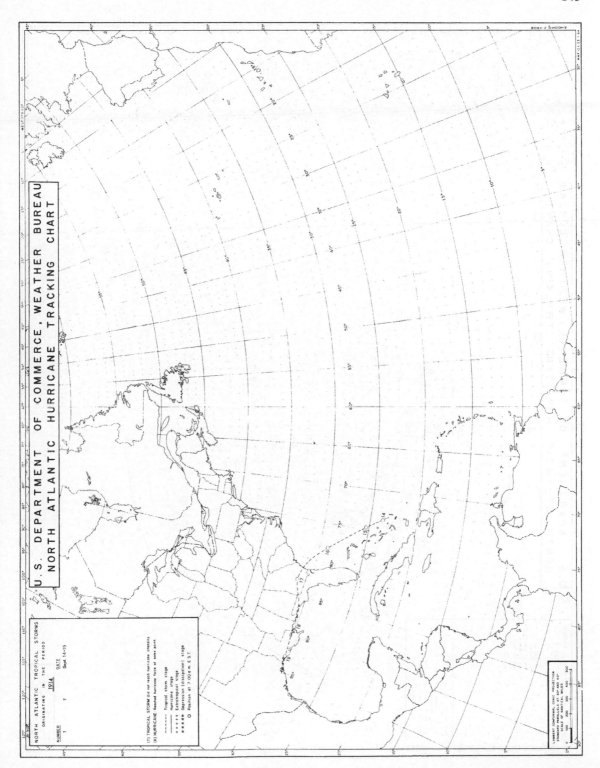

U. S. DEPARTMENT OF COMMERCE, WEATHER BUREAU
NORTH ATLANTIC HURRICANE TRACKING CHART

NORTH ATLANTIC TROPICAL STORMS
ORIGINATING IN THE PERIOD
1914

| NUMBER | DATE |
| 1 | Sept. 14-19 |

(T) TROPICAL STORM Did not reach hurricane intensity
(H) HURRICANE Reached hurricane force at some point

- - - - - - Tropical storm stage
————— Hurricane stage
+ + + + + Extratropical stage
✱ ✱ ✱ ✱ Depression (dissipation) stage
○ Position at 7:00 a.m. E.S.T.

LAMBERT CONFORMAL CONIC PROJECTION
STANDARD PARALLELS AT 30° AND 60°
SCALE OF NAUTICAL MILES

146

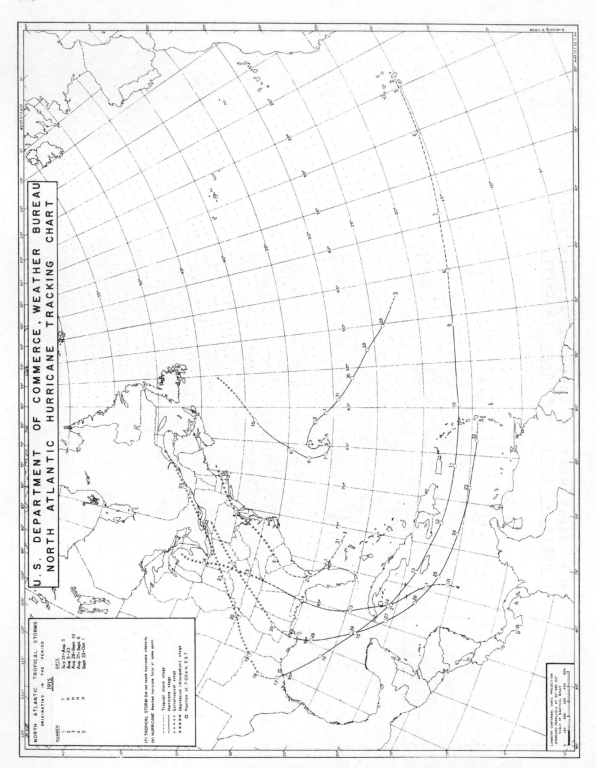

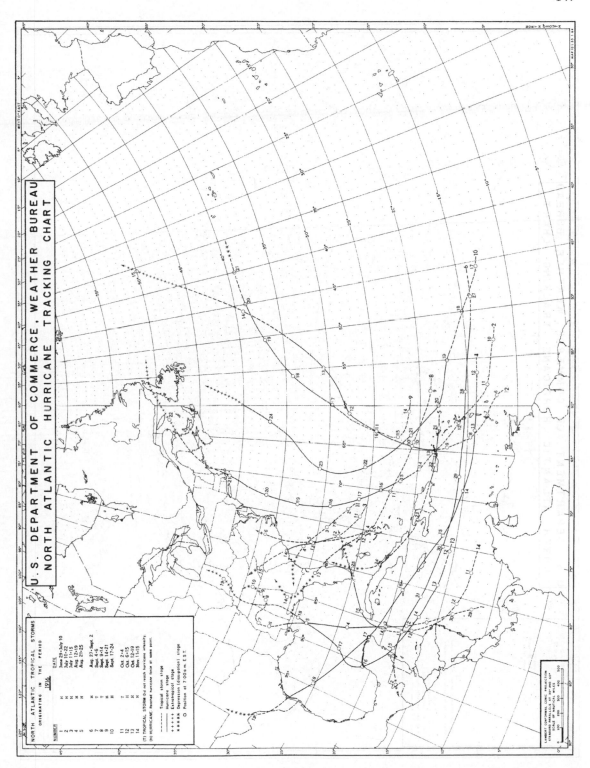

NORTH ATLANTIC TROPICAL STORMS
ORIGINATING IN THE PERIOD
1916

U. S. DEPARTMENT OF COMMERCE, WEATHER BUREAU
NORTH ATLANTIC HURRICANE TRACKING CHART

NUMBER		DATE
1	H	June 29-July 10
2	H	July 10-22
3	H	July 11-15
4	H	Aug. 12-19
5	H	Aug. 21-25
6	H	Aug. 27-Sept. 2
7	T	Sept. 4-6
8	T	Sept. 9-14
9	H	Sept. 14-21
10	H	Sept. 17-24
11	T	Oct. 2-4
12	H	Oct. 6-15
13	H	Oct. 12-19
14	H	Nov. 11-15

(T) TROPICAL STORM: Did not reach hurricane force at some point.

(H) HURRICANE: Reached hurricane force at some point.

- - - - - Tropical storm stage
———— Hurricane stage
+ + + + + Extratropical stage
× × × × × Depression (dissipation) stage
O Position at 7:00 a.m. E.S.T.

LAMBERT CONFORMAL CONIC PROJECTION
STANDARD PARALLELS AT 30° AND 60°
SCALE OF NAUTICAL MILES

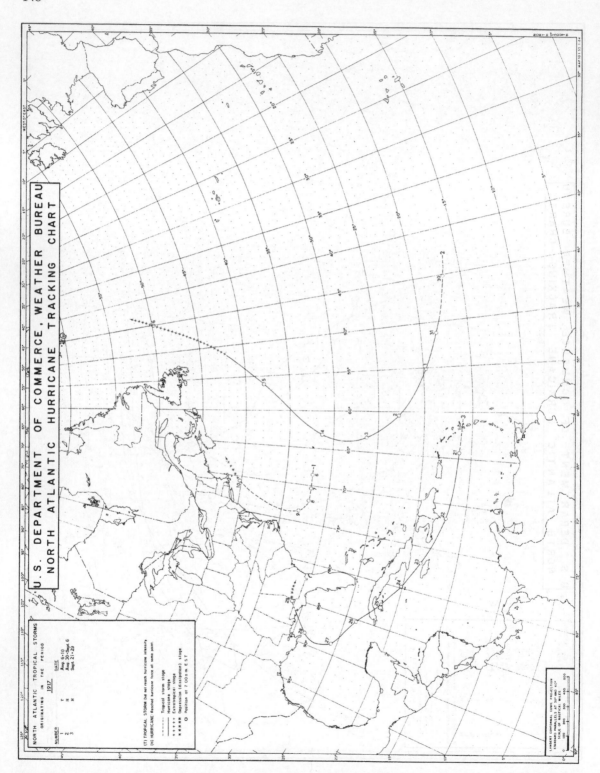

NORTH ATLANTIC TROPICAL STORMS
ORIGINATING IN THE PERIOD
1917

NUMBER		DATE
1	T	Aug. 6-10
2	H	Aug. 30-Sept. 6
3	H	Sept. 21-29

(T) TROPICAL STORM Did not reach hurricane intensity
(H) HURRICANE Reached hurricane force at some point

- - - - - Tropical storm stage
——— Hurricane stage
+ + + + + Extratropical stage
✱✱✱✱✱ Depression (dissipation) stage
O Position at 7:00 a.m. E.S.T.

U.S. DEPARTMENT OF COMMERCE, WEATHER BUREAU
NORTH ATLANTIC HURRICANE TRACKING CHART

LAMBERT CONFORMAL CONIC PROJECTION
STANDARD PARALLELS AT 30° AND 60°
SCALE OF NAUTICAL MILES

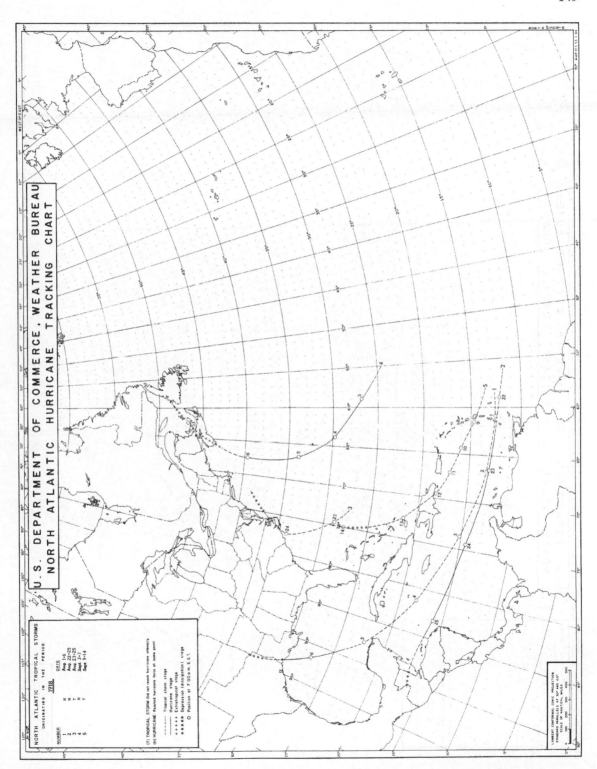

U. S. DEPARTMENT OF COMMERCE, WEATHER BUREAU
NORTH ATLANTIC HURRICANE TRACKING CHART

NORTH ATLANTIC TROPICAL STORMS
ORIGINATING IN THE PERIOD
1918

NUMBER		DATE
1	H	Aug. 1-6
2	T	Aug. 22-25
3	H	Aug. 23-25
4	H	Sept. 3-7
5	T	Sept. 9-14

(T) TROPICAL STORM: Did not reach hurricane intensity
(H) HURRICANE: Reached hurricane force at some point

——— Tropical storm stage
——— Hurricane stage
+++++ Extratropical stage
××××× Depression (dissipation) stage
O Position at 7:00 a.m. E.S.T.

LAMBERT CONFORMAL CONIC PROJECTION
STANDARD PARALLELS AT 30° AND 60°
SCALE OF NAUTICAL MILES

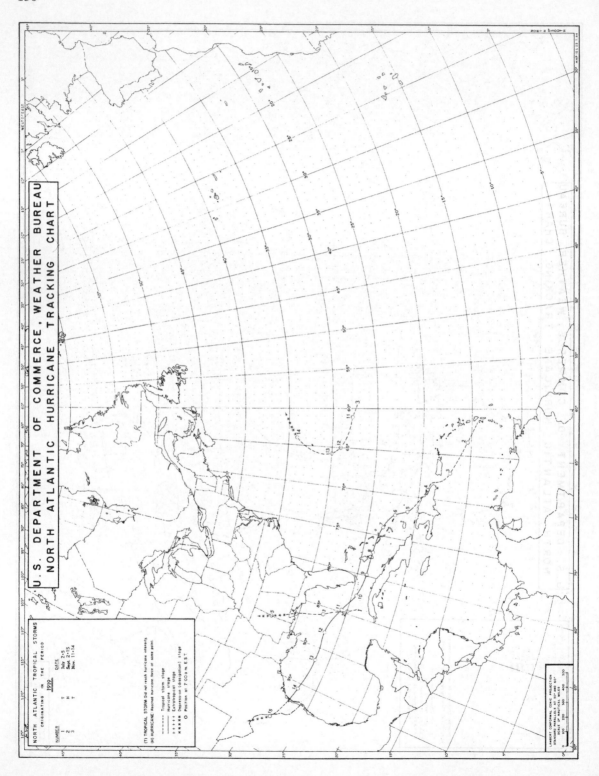

U.S. DEPARTMENT OF COMMERCE, WEATHER BUREAU
NORTH ATLANTIC HURRICANE TRACKING CHART

NORTH ATLANTIC TROPICAL STORMS
ORIGINATING IN THE PERIOD
1919

NUMBER		DATE
1	T	July 2-15
2	H	Sept. 2-15
3	T	Nov. 11-14

(T) TROPICAL STORM: Did not reach hurricane intensity.
(H) HURRICANE: Reached hurricane force at some point.

—————— Tropical storm stage
———————— Hurricane stage
+ + + + + Extratropical stage
× × × × × Depression (dissipation) stage
O Position at 7:00 a.m. E.S.T.

LAMBERT CONFORMAL CONIC PROJECTION
STANDARD PARALLELS AT 30° AND 60°
SCALE OF NAUTICAL MILES

151

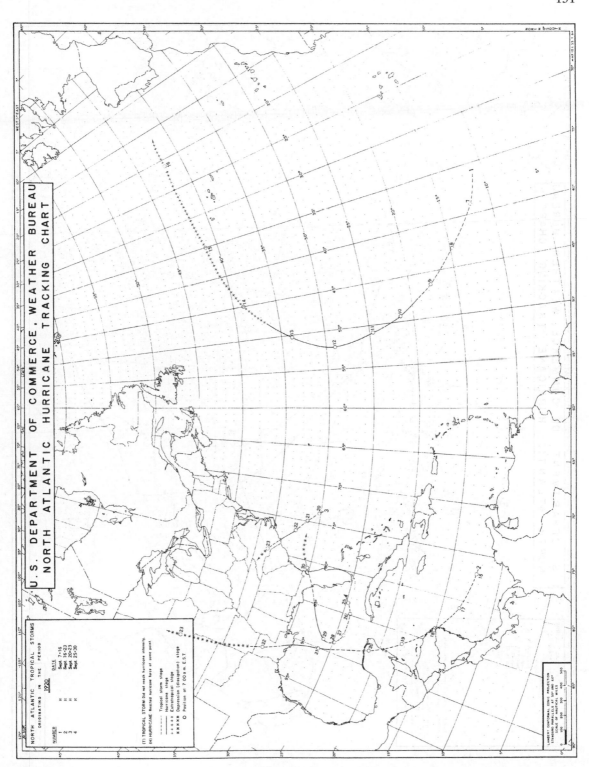

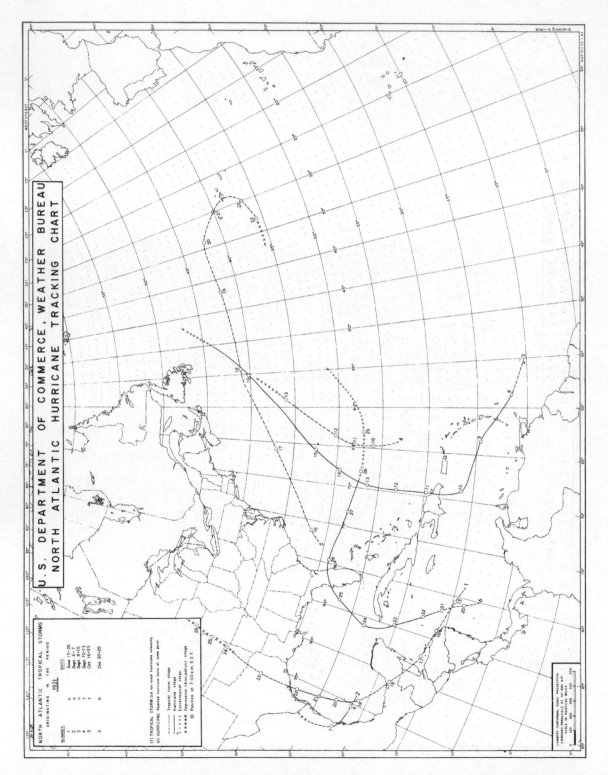

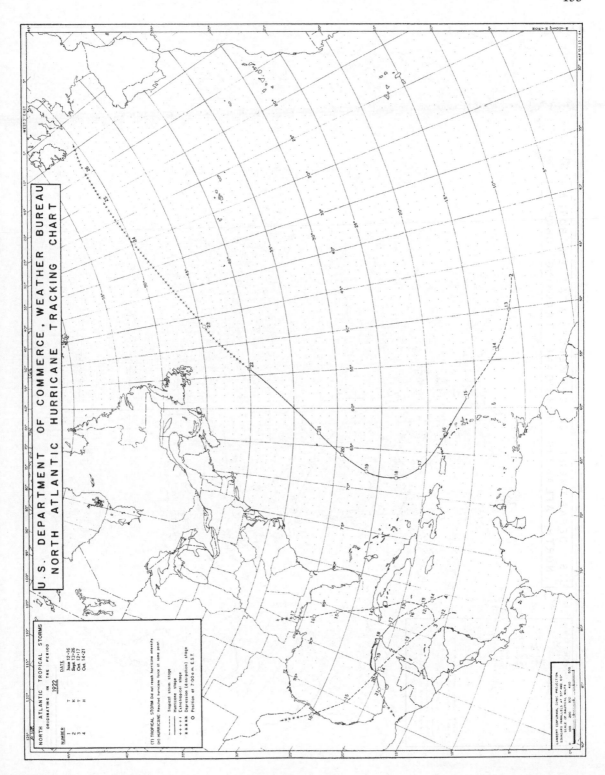

154

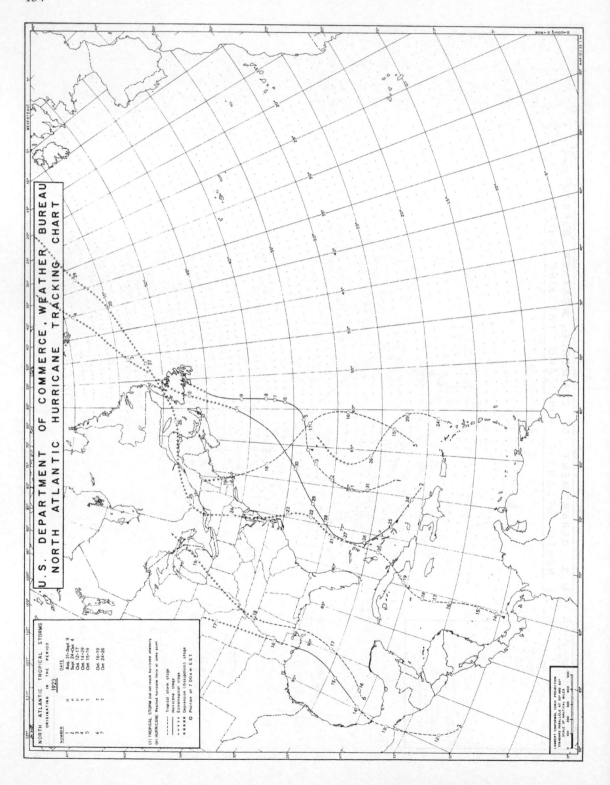

U.S. DEPARTMENT OF COMMERCE, WEATHER BUREAU
NORTH ATLANTIC HURRICANE TRACKING CHART

NORTH ATLANTIC TROPICAL STORMS
ORIGINATING IN THE PERIOD
1923

NUMBER		DATE
1	H	Aug. 31–Sept. 9
2	H	Sept. 24–Oct. 4
3	H T	Oct. 12–17
4	T	Oct. 14–29
5	T	Oct. 15–19
6	T H	Oct. 16–19
7	T H	Oct. 24–26

(T) TROPICAL STORM that did not reach hurricane intensity
(H) HURRICANE Reached hurricane force at some point

- - - - - - Tropical storm stage
+ + + + + Hurricane stage
+ + + + + Extratropical stage
× × × × × Depression (dissipation) stage
○ Position at 7:00 a.m. E.S.T.

LAMBERT CONFORMAL CONIC PROJECTION
STANDARD PARALLELS AT 30° AND 60°
SCALE OF NAUTICAL MILES

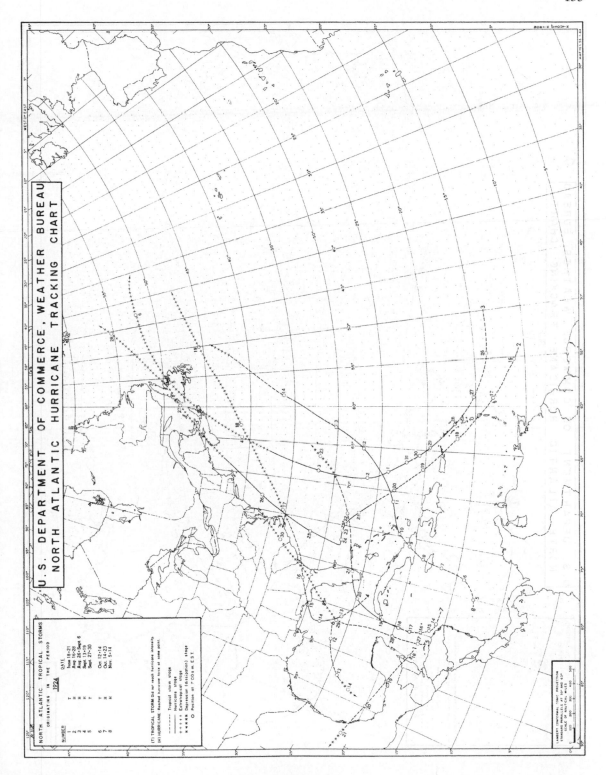

U.S. DEPARTMENT OF COMMERCE, WEATHER BUREAU
NORTH ATLANTIC HURRICANE TRACKING CHART

NORTH ATLANTIC TROPICAL STORMS
ORIGINATING IN THE PERIOD
1924

156

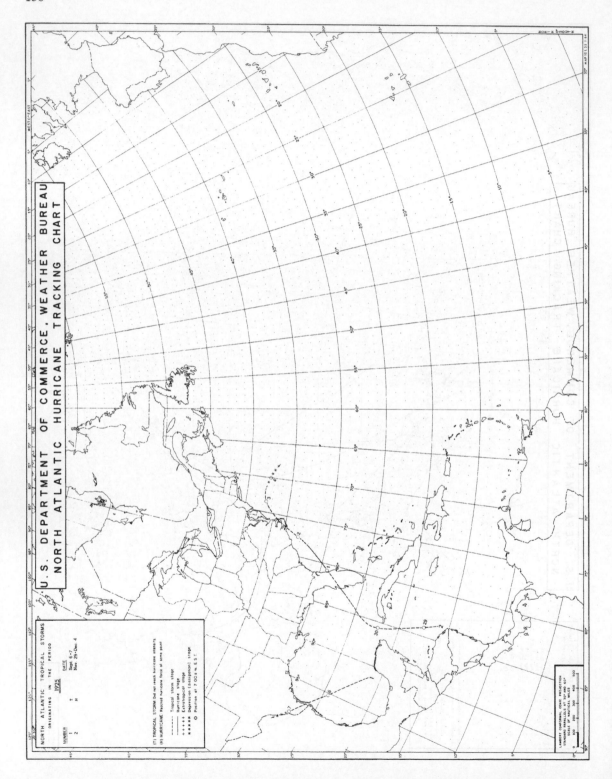

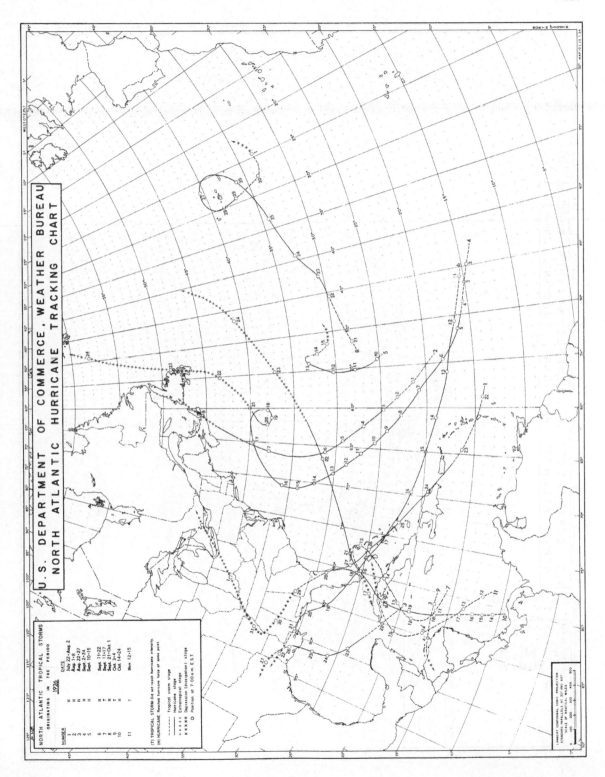

U.S. DEPARTMENT OF COMMERCE, WEATHER BUREAU
NORTH ATLANTIC HURRICANE TRACKING CHART

NORTH ATLANTIC TROPICAL STORMS
ORIGINATING IN THE PERIOD
1926

NUMBER	DATE
1 H	July 22-Aug 2
2 H	Aug 18
3 H	Aug 22-27
4 H	Sept. 2-24
5 H	Sept. 10-15
6 H T	Sept. 11-22
7 H T	Sept. 11-17
8 H T	Sept. 21-Oct. 1
9 T	Oct. 3-4
10 H	Oct. 14-24
11 T	Nov. 12-15

(T) TROPICAL STORM-Did not reach hurricane intensity.
(H) HURRICANE-Reached hurricane force at some point.

- - - - - Tropical storm stage
+ + + + + Hurricane stage
- - - - - Extratropical stage
× × × × × Depression (extratropical) stage
○ Position at 7:00 a.m. E.S.T.

LAMBERT CONFORMAL CONIC PROJECTION
STANDARD PARALLELS AT 30° AND 60°
SCALE OF NAUTICAL MILES

158

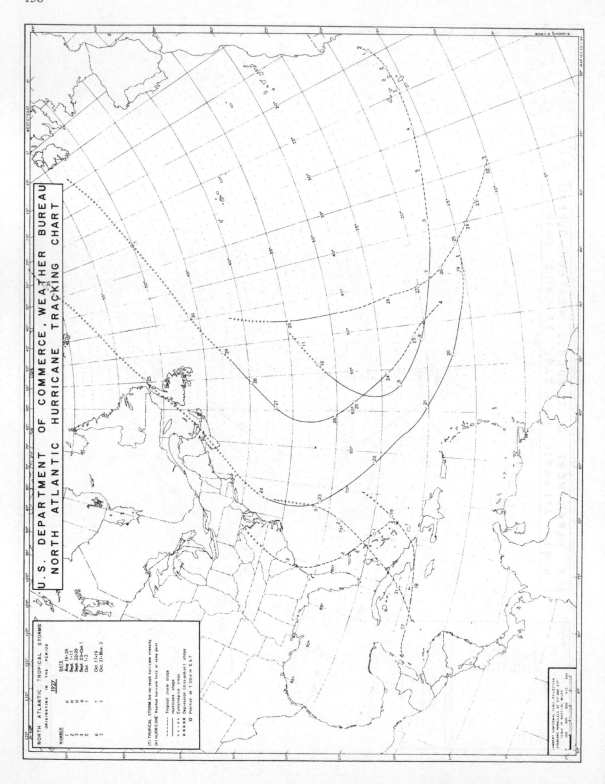

159

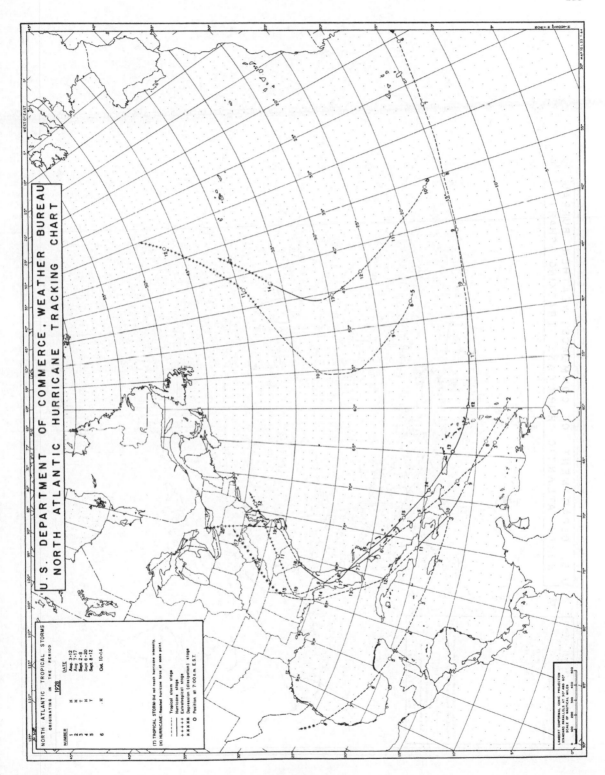

160

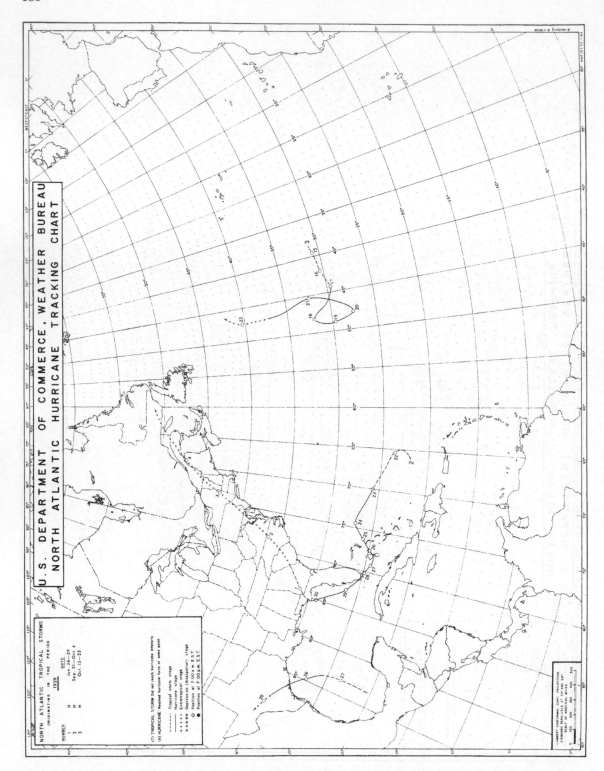

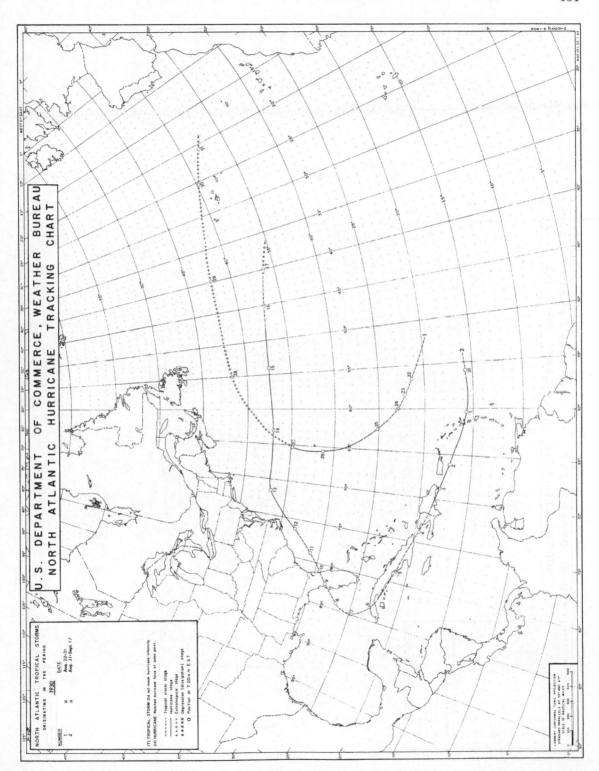

U.S. DEPARTMENT OF COMMERCE, WEATHER BUREAU
NORTH ATLANTIC HURRICANE TRACKING CHART

NORTH ATLANTIC TROPICAL STORMS
ORIGINATING IN THE PERIOD
1930

NUMBER		DATE
1	H	Aug. 22-31
2	H	Aug. 31-Sept. 17

(T) TROPICAL STORM Did not reach hurricane intensity.
(H) HURRICANE Reached hurricane force at some point.

- - - - - Tropical storm stage
+ + + + + Hurricane stage
+ + + + + Extratropical stage
× × × × × Depression (dissipation) stage
O Position at 7.00 a.m. E.S.T.

LAMBERT CONFORMAL CONIC PROJECTION
STANDARD PARALLELS AT 59° AND 42°
SCALE OF NAUTICAL MILES

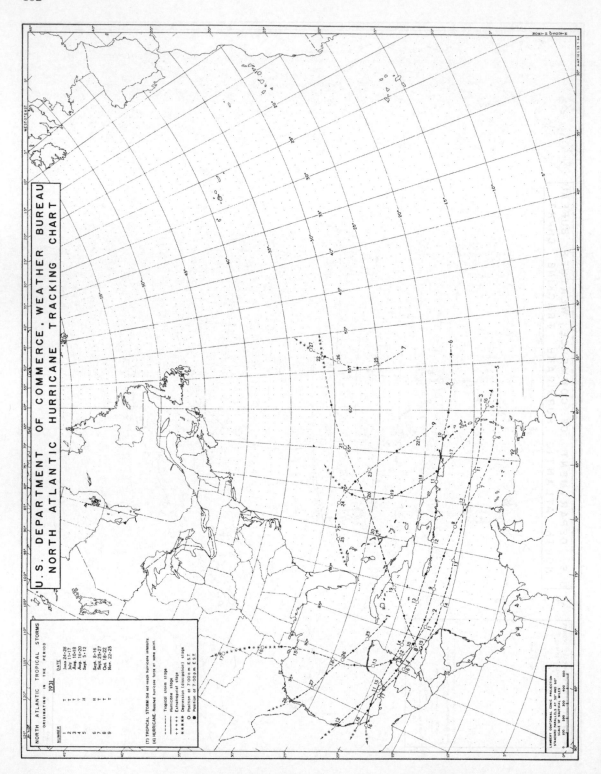

162

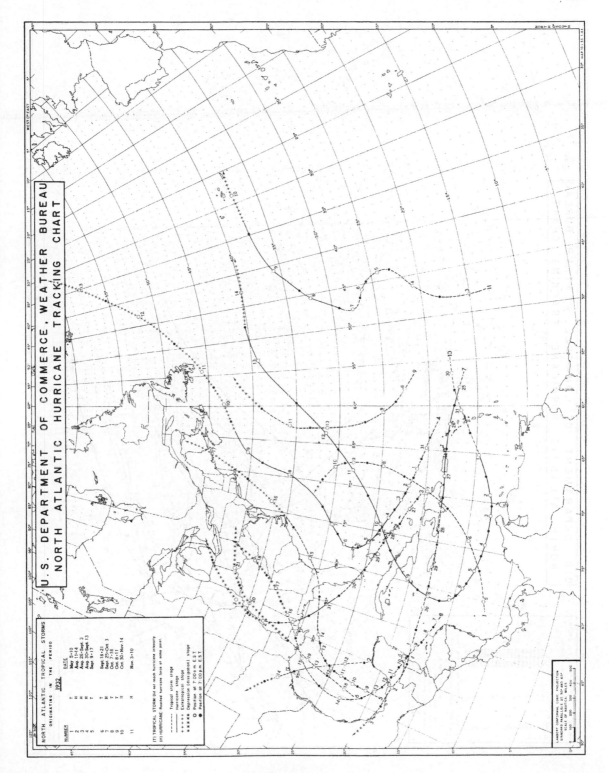

NORTH ATLANTIC TROPICAL STORMS
ORIGINATING IN THE PERIOD
1932

U.S. DEPARTMENT OF COMMERCE, WEATHER BUREAU
NORTH ATLANTIC HURRICANE TRACKING CHART

164

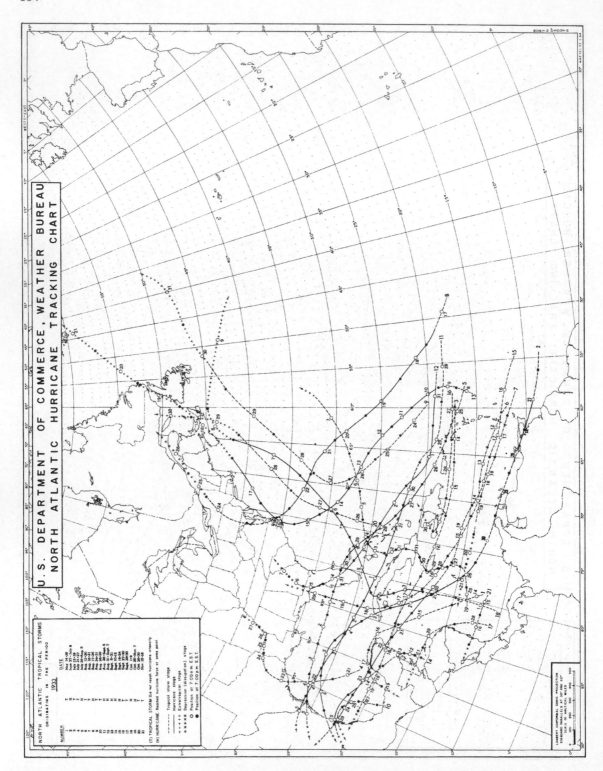

165

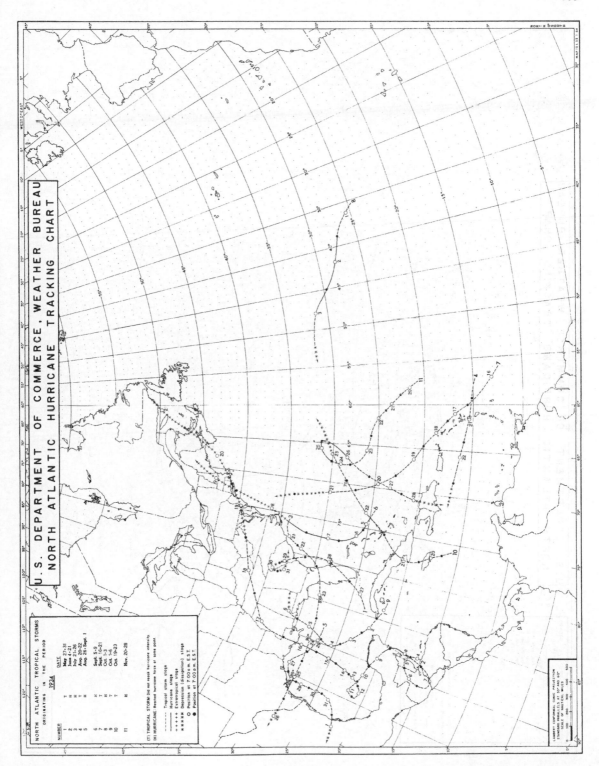

NORTH ATLANTIC TROPICAL STORMS
ORIGINATING IN THE PERIOD
1934

U.S. DEPARTMENT OF COMMERCE, WEATHER BUREAU
NORTH ATLANTIC HURRICANE TRACKING CHART

166

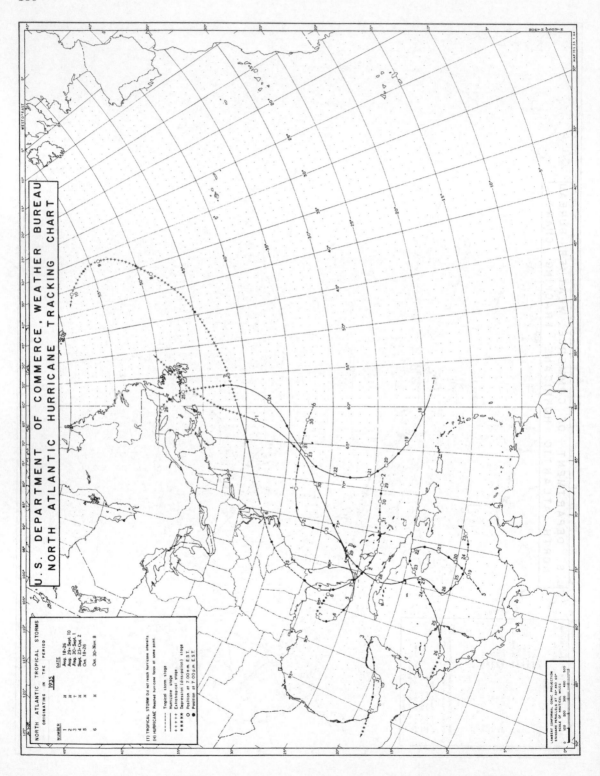

U.S. DEPARTMENT OF COMMERCE, WEATHER BUREAU
NORTH ATLANTIC HURRICANE TRACKING CHART

NORTH ATLANTIC TROPICAL STORMS
ORIGINATING IN THE PERIOD
1925

NUMBER		DATE
1	H	Aug. 18-26
2	H	Aug. 29-Sept. 10
3	T	Aug. 30-Sept. 1
4	H	Sept. 23-Oct. 2
5	H	Oct. 18-26
6	H	Oct. 30-Nov. 8

(T) TROPICAL STORM Did not reach hurricane intensity.

(H) HURRICANE Reached hurricane force at some point.

————— Tropical storm stage
+ + + + + Hurricane stage
* * * * * Extratropical stage
●●●●● Depression (dissipation) stage
○ Position at 7:00 a.m. E.S.T.
● Position at 7:00 p.m. E.S.T.

LAMBERT CONFORMAL CONIC PROJECTION
STANDARD PARALLELS AT 30° AND 60°
SCALE OF NAUTICAL MILES

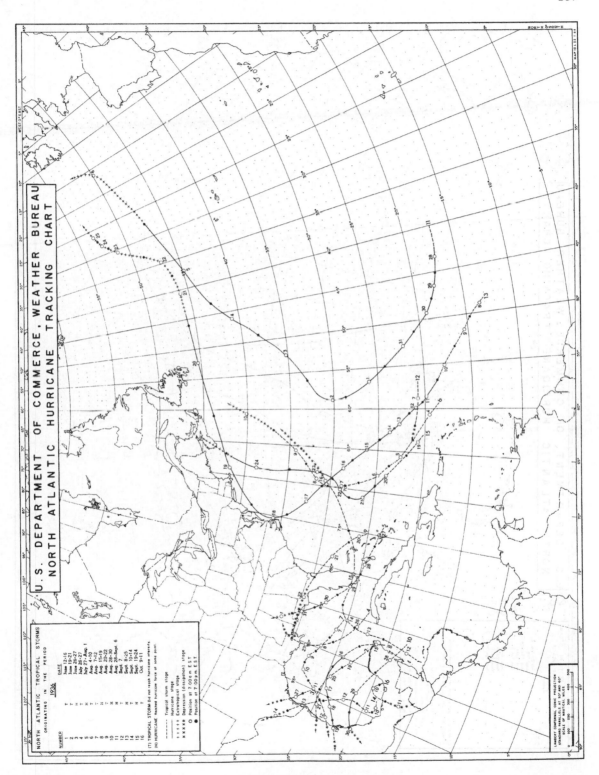

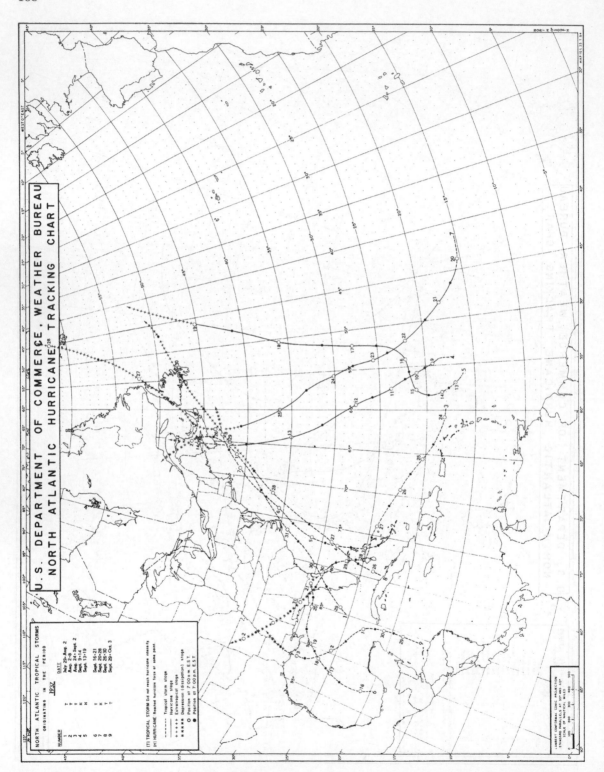

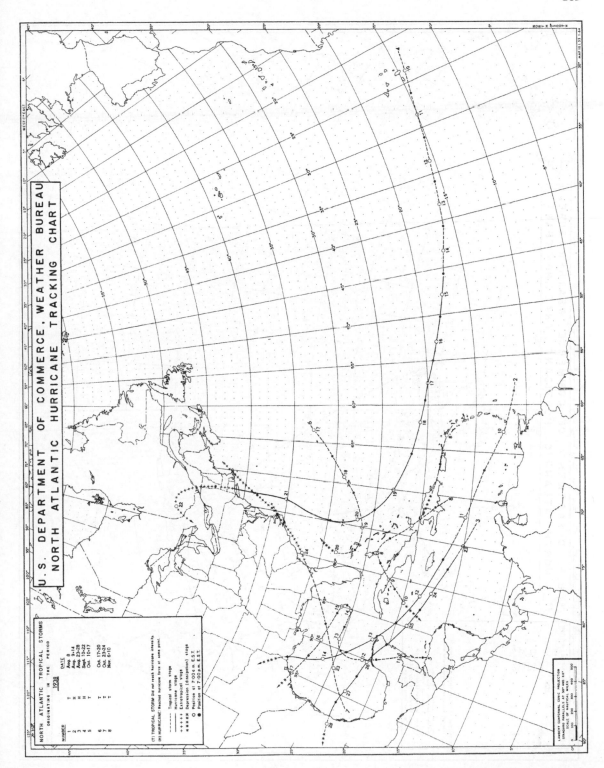

U.S. DEPARTMENT OF COMMERCE, WEATHER BUREAU
NORTH ATLANTIC HURRICANE TRACKING CHART

NORTH ATLANTIC TROPICAL STORMS
ORIGINATING IN THE PERIOD
1938

NUMBER		DATE
1	T	Aug. 8
2	H	Aug. 9-14
3	H	Aug. 23-28
4	H	Sep. 10-22
5	T	Oct. 10-17
6	T	Oct. 17-20
7	T	Oct. 23-24
8	T	Nov. 6-10

(T) TROPICAL STORM Did not reach hurricane intensity.
(H) HURRICANE Reached hurricane force at some point.

───── Tropical storm stage
∘∘∘∘∘ Hurricane stage
+ + + + Extratropical stage
× × × × Depression (dissipation) stage
○ Position at 7:00 a.m. E.S.T.
● Position at 7:00 p.m. E.S.T.

LAMBERT CONFORMAL CONIC PROJECTION
STANDARD PARALLELS AT 36° AND 60°
SCALE OF NAUTICAL MILES

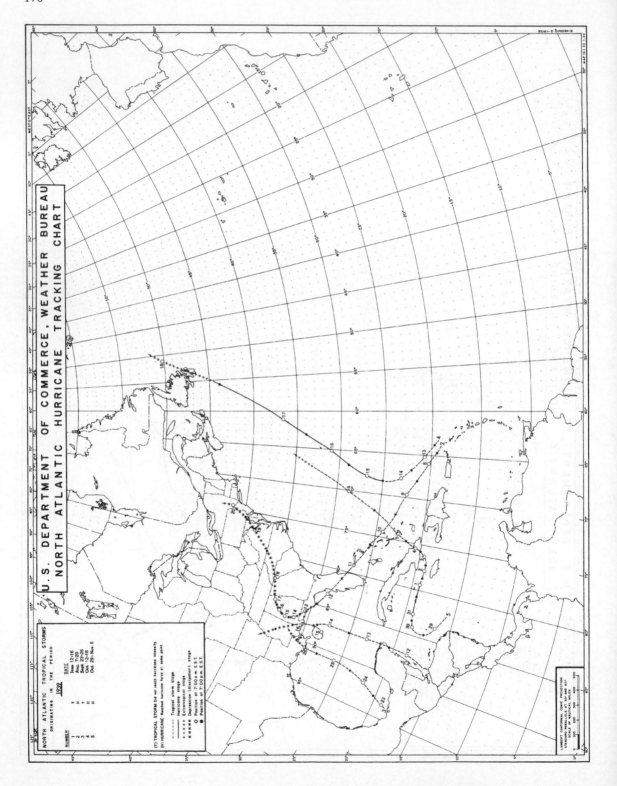

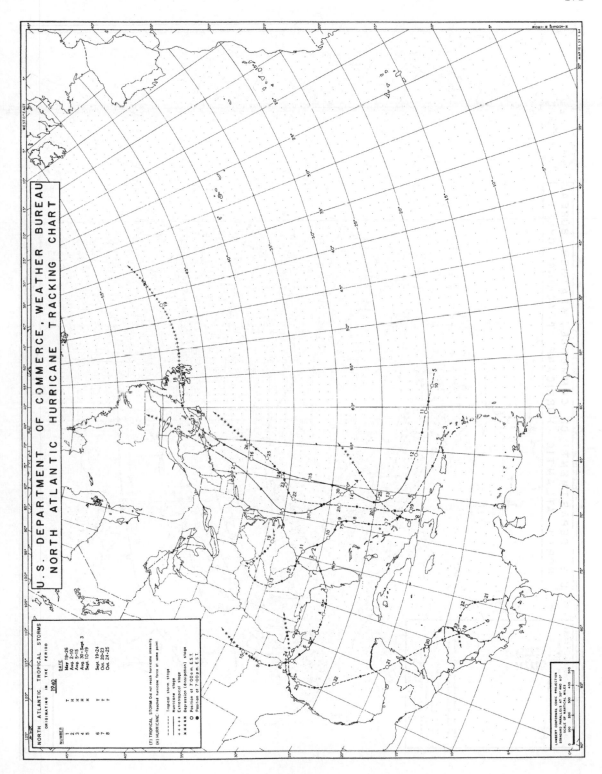

NORTH ATLANTIC TROPICAL STORMS
ORIGINATING IN THE PERIOD
1940

U. S. DEPARTMENT OF COMMERCE, WEATHER BUREAU
NORTH ATLANTIC HURRICANE TRACKING CHART

172

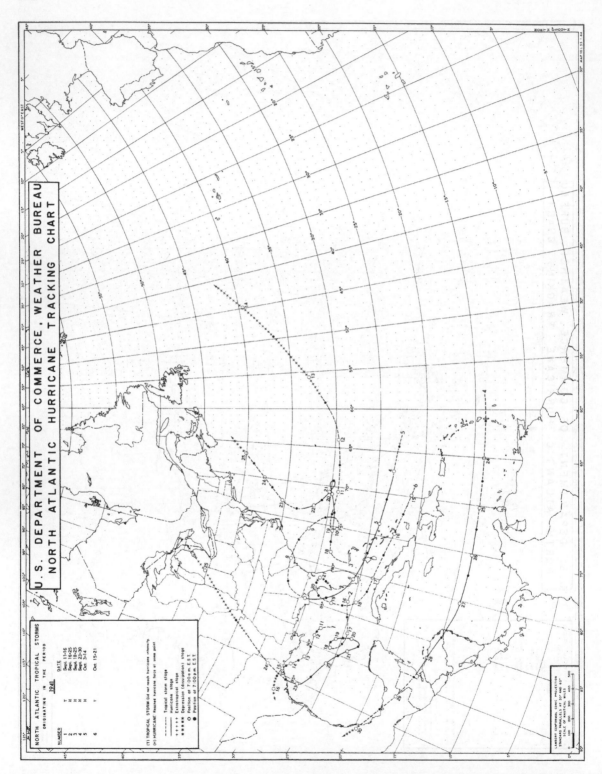

173

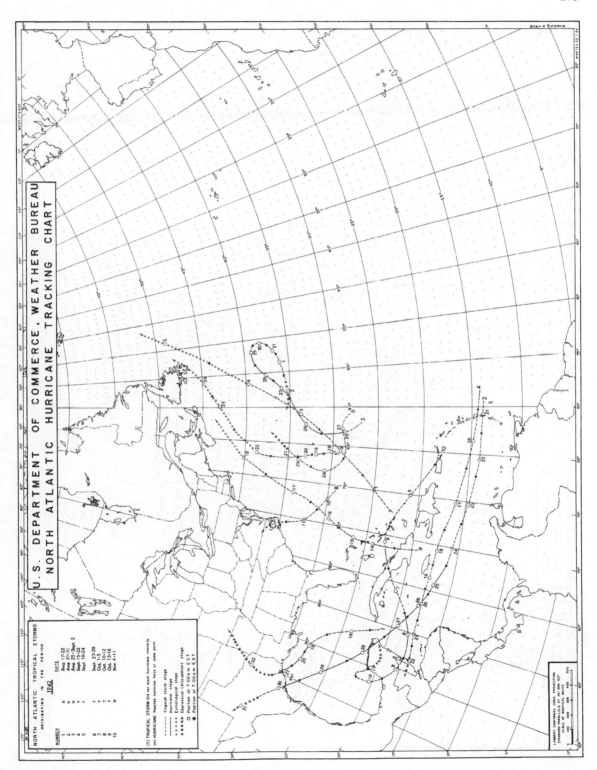

174

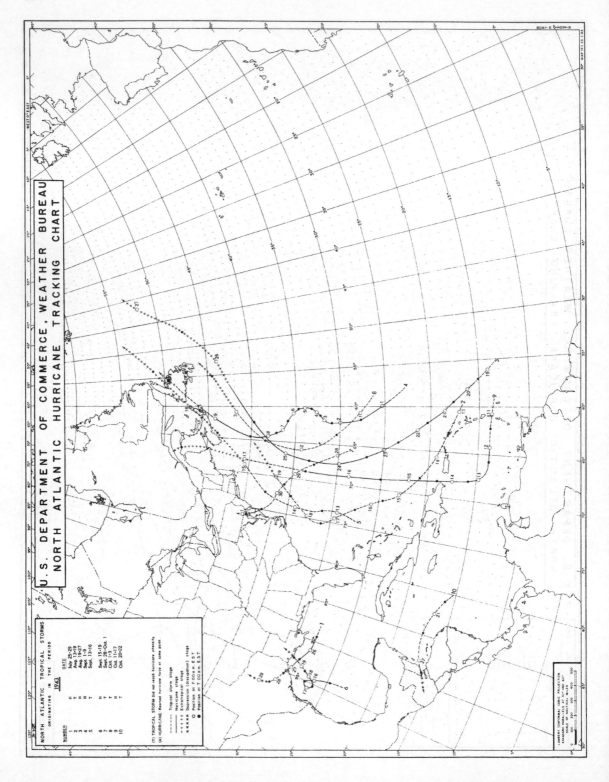

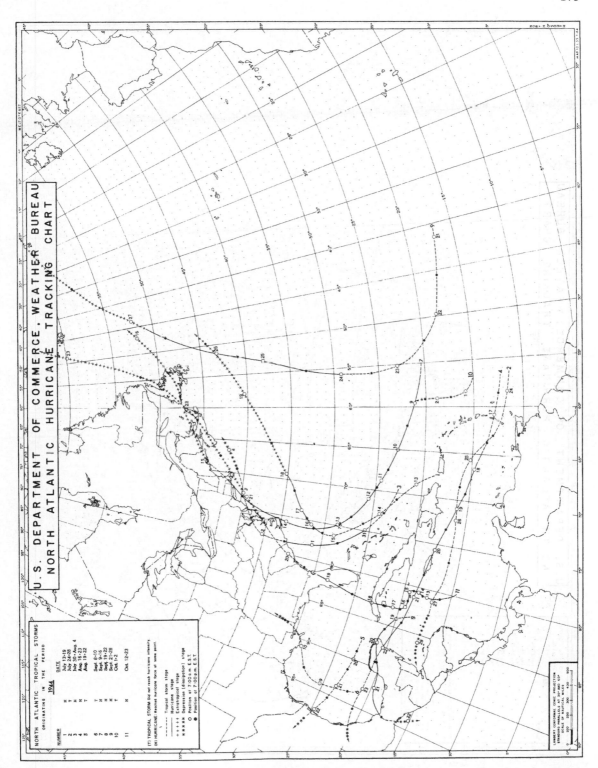

NORTH ATLANTIC TROPICAL STORMS
ORIGINATING IN THE PERIOD

1944

NUMBER		DATE
1	H	July 13-19
2	T	July 24-28
3	H	July 30-Aug. 4
4	H	Aug. 16-23
5	T	Aug. 19-22
6	T	Sept. 8-10
7	H	Sept. 9-16
8	H	Sept. 19-22
9	T	Sept. 21-28
10	H	Oct. 1-2
11	H	Oct. 12-23

(T) TROPICAL STORM: Did not reach hurricane intensity.
(H) HURRICANE: Reached hurricane force at some point.

— — — — Tropical storm stage
———— Hurricane stage
+ + + + Extratropical stage
× × × × Depression (dissipation) stage
○ Position at 7:00 a.m. E.S.T.
● Position at 7:00 p.m. E.S.T.

U. S. DEPARTMENT OF COMMERCE, WEATHER BUREAU
NORTH ATLANTIC HURRICANE TRACKING CHART

LAMBERT CONFORMAL CONIC PROJECTION
STANDARD PARALLELS AT 30° AND 60°
SCALE OF NAUTICAL MILES

176

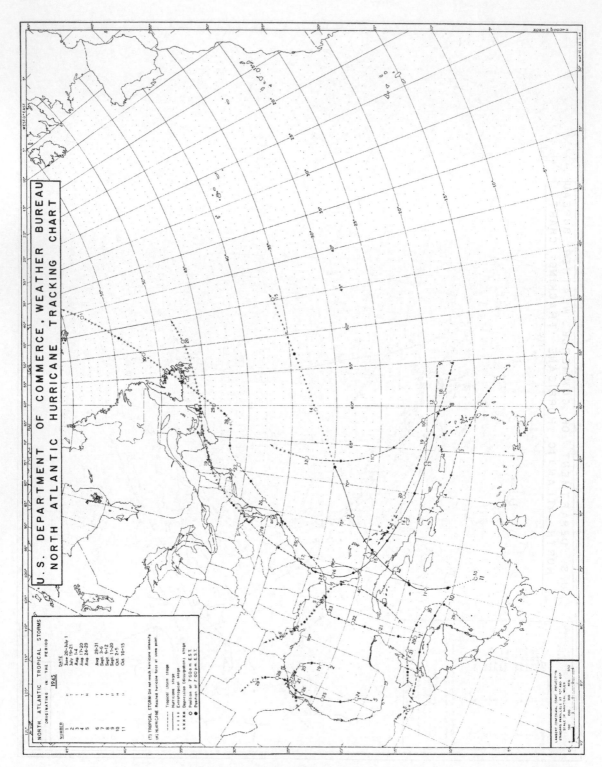

U.S. DEPARTMENT OF COMMERCE, WEATHER BUREAU
NORTH ATLANTIC HURRICANE TRACKING CHART

NORTH ATLANTIC TROPICAL STORMS

ORIGINATING IN THE PERIOD

1945

NUMBER	DATE
1	June 20–July 1
2	July 19–21
3	Aug. 1–4
4	Aug. 17–20
5	Aug. 24–29
6	Aug. 25–31
7	Sept. 3–6
8	Sept. 9–12
9	Sept. 11–20
10	Oct. 2–5
11	Oct. 10–15

(T) TROPICAL STORM Did not reach hurricane intensity

(H) HURRICANE Reached hurricane force at some point

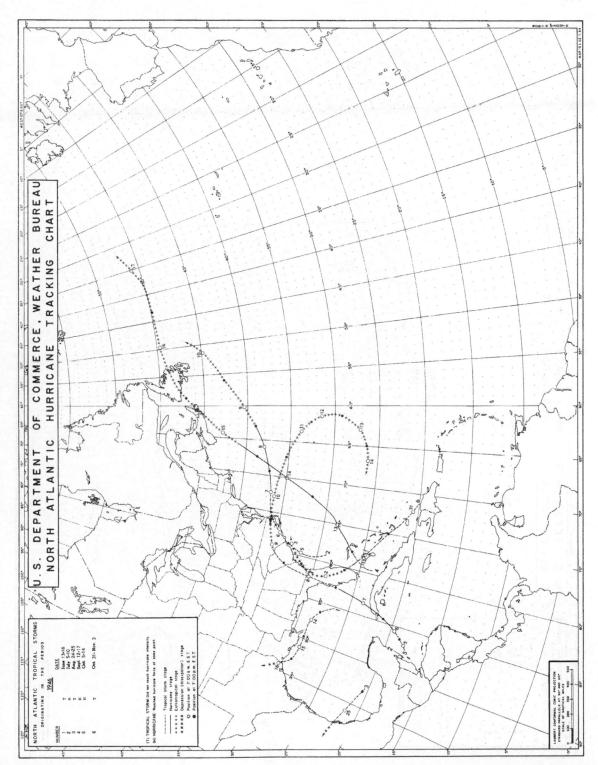

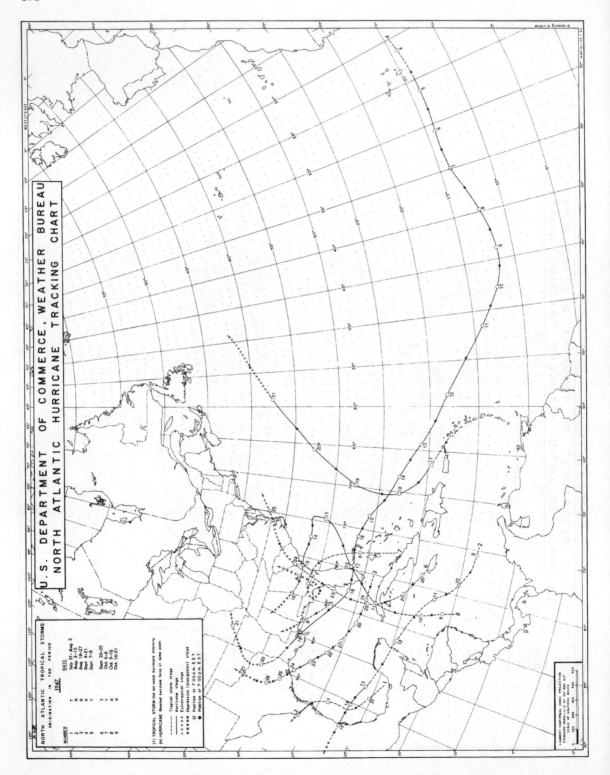

U.S. DEPARTMENT OF COMMERCE, WEATHER BUREAU
NORTH ATLANTIC HURRICANE TRACKING CHART

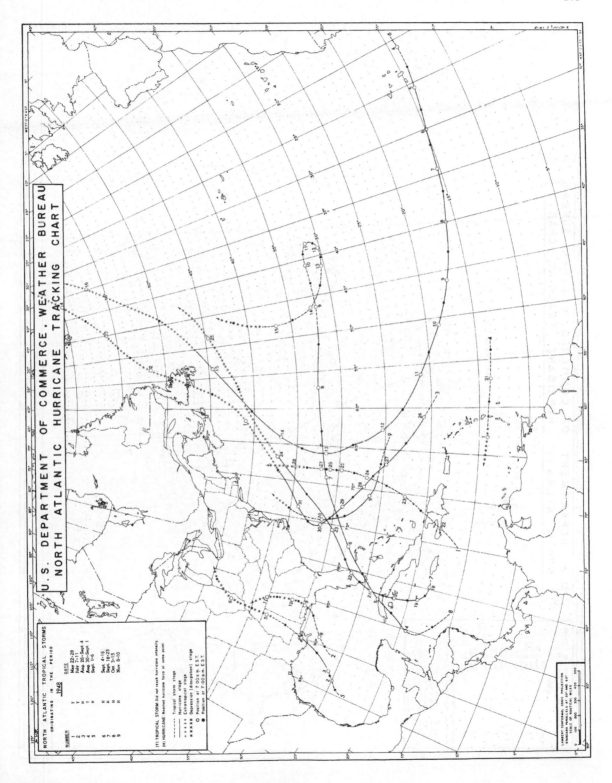

U.S. DEPARTMENT OF COMMERCE, WEATHER BUREAU

NORTH ATLANTIC HURRICANE TRACKING CHART

NORTH ATLANTIC TROPICAL STORMS
ORIGINATING IN THE PERIOD
1948

NUMBER		DATE
1	T	May 22-28
2	T	July 7-11
3	H	Aug. 26-Sept. 4
4	T	Aug. 30-Sept. 1
5	H	Sept. 1-6
6	H	Sept. 4-16
7	H	Sept. 18-25
8	H	Oct. 3-15
9	H	Nov. 8-10

(T) TROPICAL STORM Did not reach hurricane intensity.
(H) HURRICANE Reached hurricane force at some point.

——————— Tropical storm stage
– – – – – Hurricane stage
+ + + + + Extratropical stage
× × × × × Depression (dissipation) stage
○ Position at 7:00 a.m. E.S.T.
● Position at 7:00 p.m. E.S.T.

LAMBERT CONFORMAL CONIC PROJECTION
STANDARD PARALLELS AT 20° AND 60°
SCALE OF NAUTICAL MILES

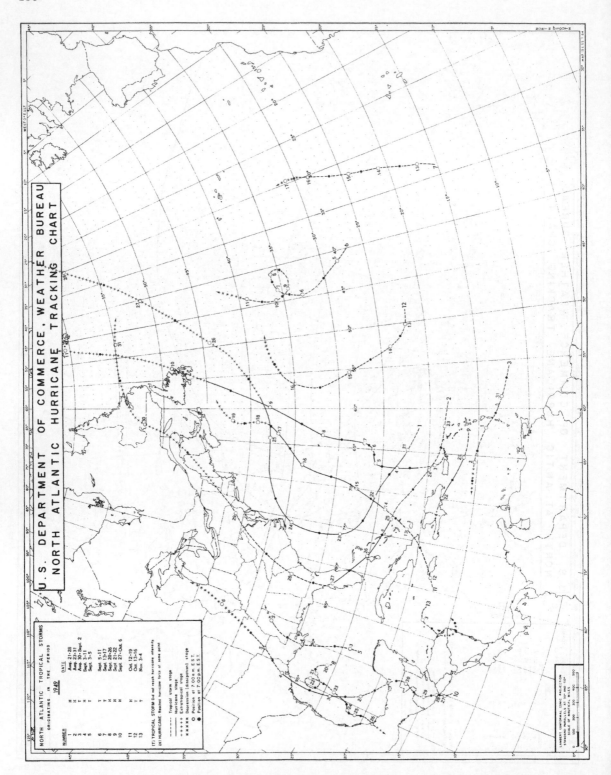

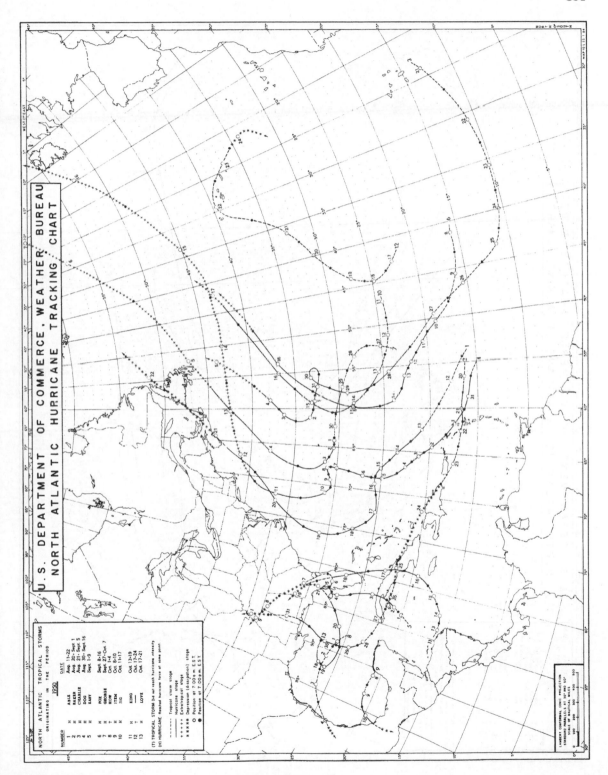

U.S. DEPARTMENT OF COMMERCE, WEATHER BUREAU

NORTH ATLANTIC HURRICANE TRACKING CHART

182

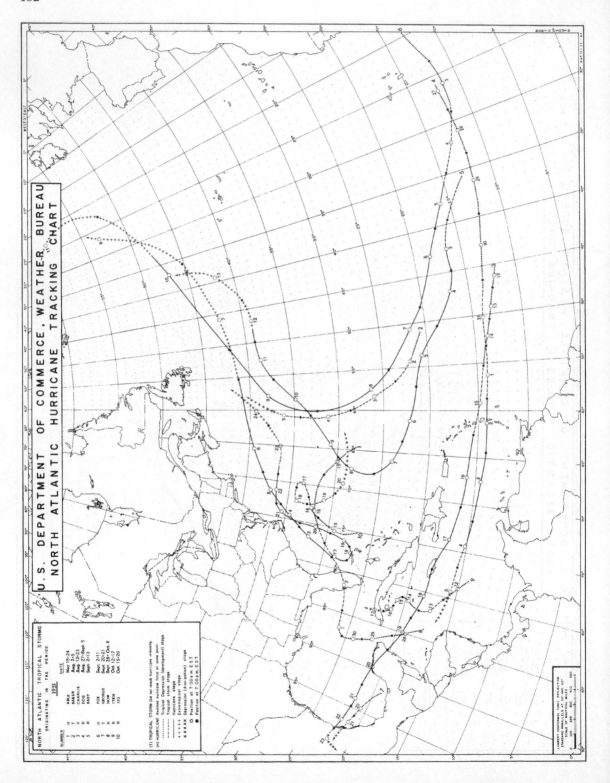

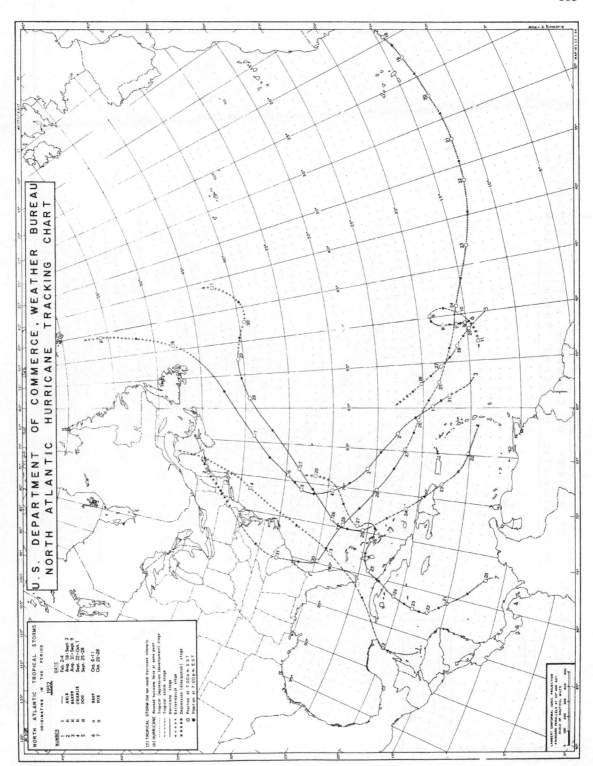

U.S. DEPARTMENT OF COMMERCE, WEATHER BUREAU
NORTH ATLANTIC HURRICANE TRACKING CHART

NORTH ATLANTIC TROPICAL STORMS
ORIGINATING IN THE PERIOD
1952

NUMBER			DATE
1	T		Feb. 2-4
2	H	ABLE	Aug. 18-Sept. 2
3	H	BAKER	Aug. 31-Sept. 9
4	H	CHARLIE	Sept. 25-Oct. 1
5	H	DOG	Sept. 25-28
6	H	EASY	Oct. 6-11
7	H	FOX	Oct. 20-28

(T) TROPICAL STORM Did not reach hurricane intensity
(H) HURRICANE Reached hurricane force at some point
·········· Tropical Depression (development) stage
———— Tropical storm stage
+++++ Hurricane stage
××××× Extratropical stage
★★★★★ Depression (dissipation) stage
○ Position at 7.00 a.m. E.S.T.
● Position at 7.00 p.m. E.S.T.

LAMBERT CONFORMAL CONIC PROJECTION
STANDARD PARALLELS AT 30° AND 60°
SCALE OF NAUTICAL MILES

184

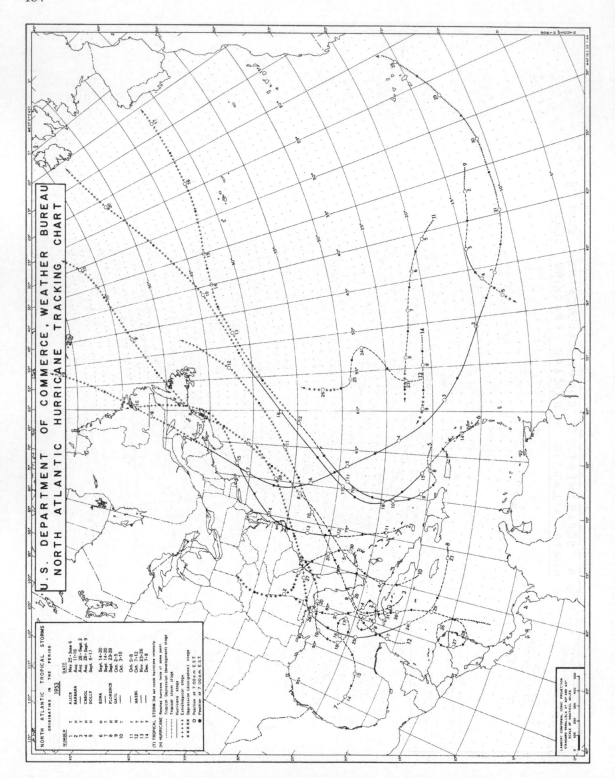

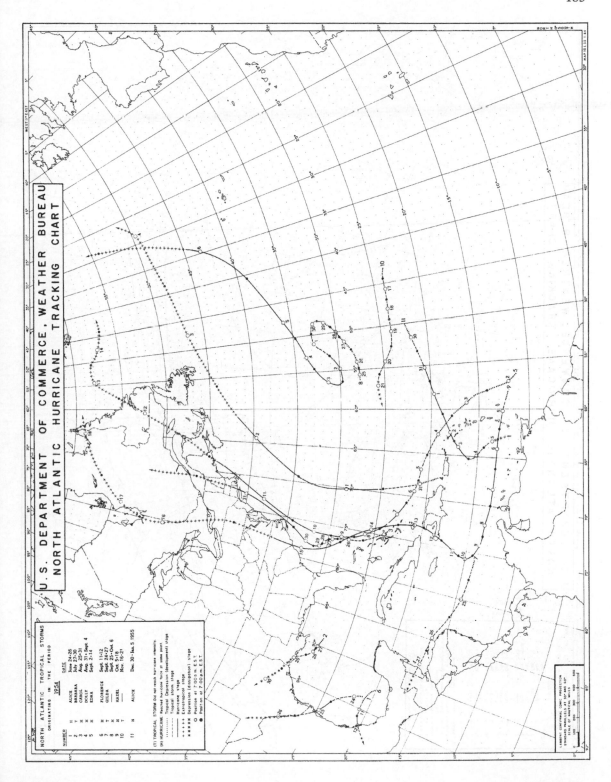

NORTH ATLANTIC TROPICAL STORMS

ORIGINATING IN THE PERIOD

1954

NUMBER		DATE
1	H	ALICE	June 24-25
2	T	BARBARA	July 27-30
3	H	CAROL	Aug. 25-31
4	H	DOLLY	Aug. 31-Sept. 4
5	H	EDNA	Sept. 2-14
6	H	FLORENCE	Sept. 11-12
7	T	GILDA	Sept. 24-27
8	H	HAZEL	Sept. 25-Oct. 6
9	H		Oct. 5-18
10	T		Nov. 16-21
11	H	ALICE	Dec. 30-Jan. 5 1955

(T) TROPICAL STORM. Did not reach hurricane intensity.

(H) HURRICANE. Reached hurricane force at same point.

········· Tropical Depression (development) stage

─────── Tropical storm stage

++++++ Hurricane stage

×××××× Extratropical (dissipation) stage

○ Position at 7:00 a.m. E.S.T.

● Position at 7:00 p.m. E.S.T.

U.S. DEPARTMENT OF COMMERCE, WEATHER BUREAU

NORTH ATLANTIC HURRICANE TRACKING CHART

186

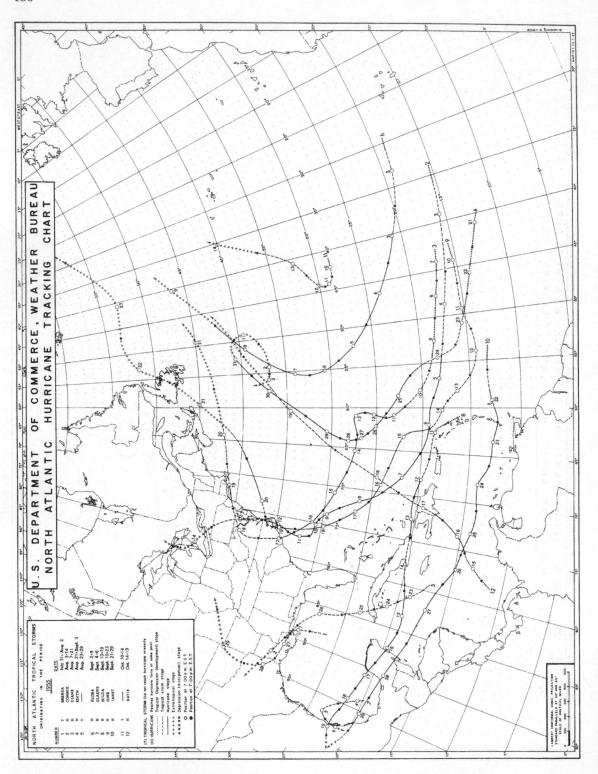

U.S. DEPARTMENT OF COMMERCE, WEATHER BUREAU
NORTH ATLANTIC HURRICANE TRACKING CHART

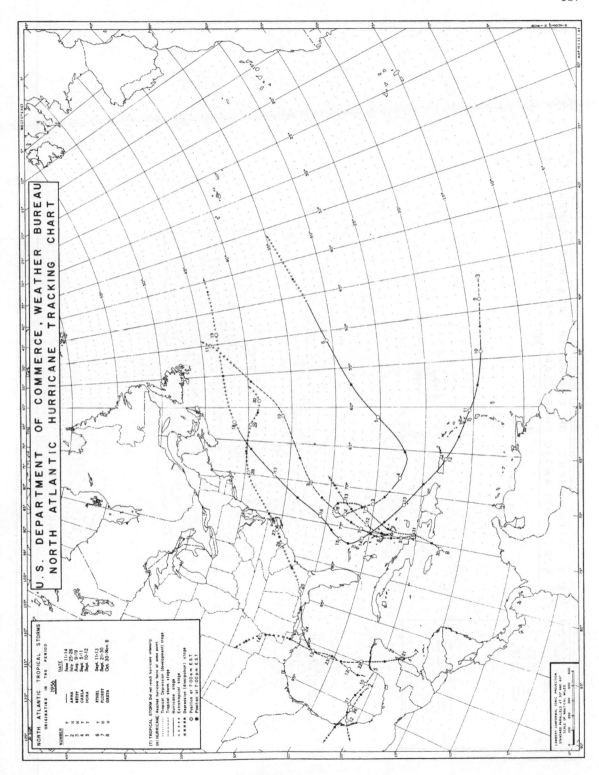

NORTH ATLANTIC TROPICAL STORMS
ORIGINATING IN THE PERIOD
1956

NUMBER			DATE
1	T	ANNA	June 11-14
2	H	BETSY	July 25-26
3	H	CARLA	Aug. 9-19
4	T	DORA	Sep. 5-11
5	T		Sept. 10-12
6	T	ETHEL	Sept. 11-13
7	H	FLOSSY	Sept. 21-30
8	H	GRETA	Oct. 30-Nov. 6

(T) TROPICAL STORM Did not reach hurricane intensity
(H) HURRICANE Reached hurricane force at some point
 ---------- Tropical Depression (development) stage
 ·········· Tropical storm stage
 ++++++ Hurricane stage
 ***** Extratropical stage
 ○ Depression (dissipation) stage
 ● Position at 7:00 a.m. E.S.T.
 ● Position at 7:00 p.m. E.S.T.

U.S. DEPARTMENT OF COMMERCE, WEATHER BUREAU
NORTH ATLANTIC HURRICANE TRACKING CHART

LAMBERT CONFORMAL CONIC PROJECTION
STANDARD PARALLELS AT 30° AND 60°
SCALE OF NAUTICAL MILES

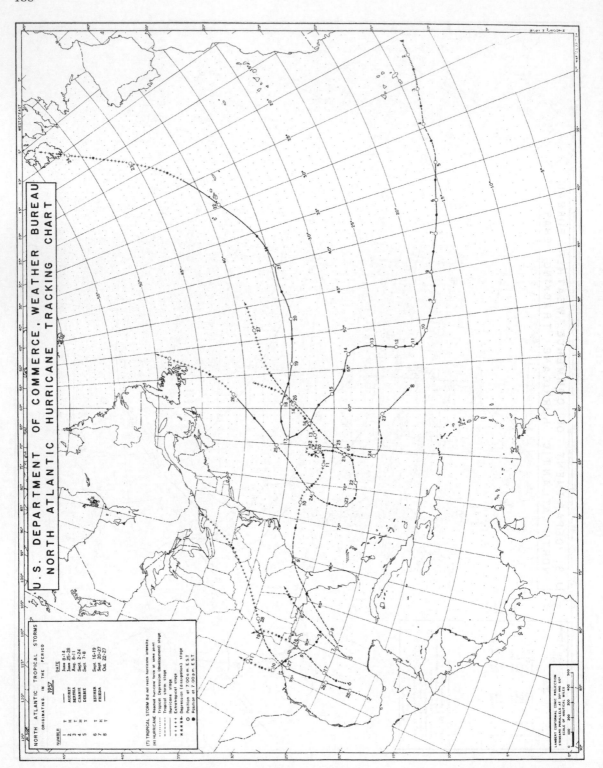

U.S. DEPARTMENT OF COMMERCE, WEATHER BUREAU
NORTH ATLANTIC HURRICANE TRACKING CHART

NORTH ATLANTIC TROPICAL STORMS
ORIGINATING IN THE PERIOD
1957

NUMBER			DATE
1	T		June 8-14
2	H	AUDREY	June 25-28
3	T	BERTHA	Aug. 8-11
4	T	CARRIE	Sept. 2-24
5	T	DEBBIE	Sept. 7-8
6	T	ESTHER	Sept. 16-19
7	H	FRIEDA	Sept. 20-27
8	T		Oct. 22-27

(T) TROPICAL STORM Did not reach hurricane intensity.
(H) HURRICANE Reached hurricane force at some point.
·········· Tropical Depression (development) stage
- - - - - Tropical storm stage
+++++ Hurricane stage
▲▲▲▲ Extratropical stage
○ Position at 7:00 a.m. E.S.T.
● Position at 7:00 p.m. E.S.T.

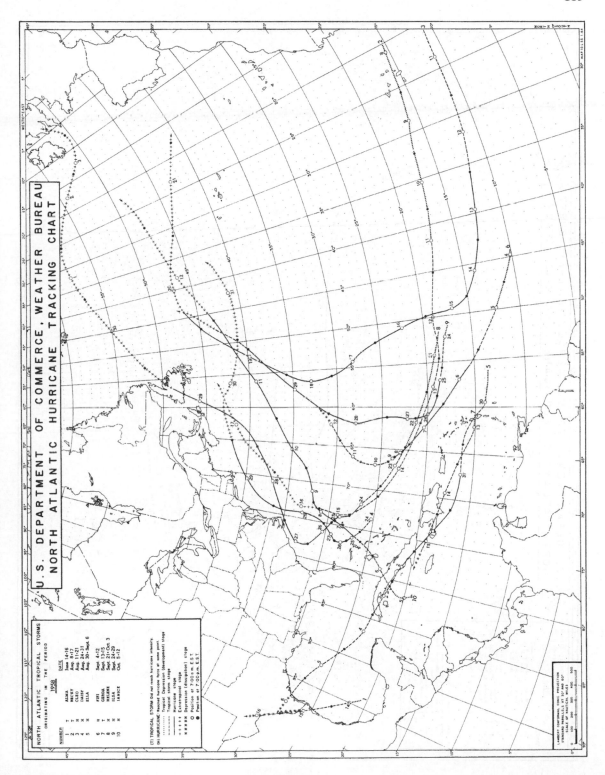

U. S. DEPARTMENT OF COMMERCE, WEATHER BUREAU
NORTH ATLANTIC HURRICANE TRACKING CHART

NORTH ATLANTIC TROPICAL STORMS
ORIGINATING IN THE PERIOD
1958

NUMBER		NAME	DATE
1	T	ALMA	June 14-16
2	T	BECKY	Aug. 8-17
3	H	CLEO	Aug. 11-21
4	H	DAISY	Aug. 24-31
5	H	ELLA	Aug. 30-Sep. 6
6	H	FIFI	Sept. 4-12
7	T	GERDA	Sept. 13-15
8	H	HELENE	Sept. 21-Oct. 3
9	H	ILSA	Sept. 24-29
10	H	JANICE	Oct. 5-12

(T) TROPICAL STORM: Did not reach hurricane intensity.
(H) HURRICANE: Reached hurricane force at some point.

.... Tropical Depression (development) stage
———— Tropical storm stage
———— Hurricane stage
+ + + + Extratropical stage
✕ ✕ ✕ ✕ Depression (dissipation) stage
○ Position at 7:00 a.m. EST
● Position at 7:00 p.m. EST

LAMBERT CONFORMAL CONIC PROJECTION
STANDARD PARALLELS AT 30° AND 60°
SCALE OF NAUTICAL MILES

190

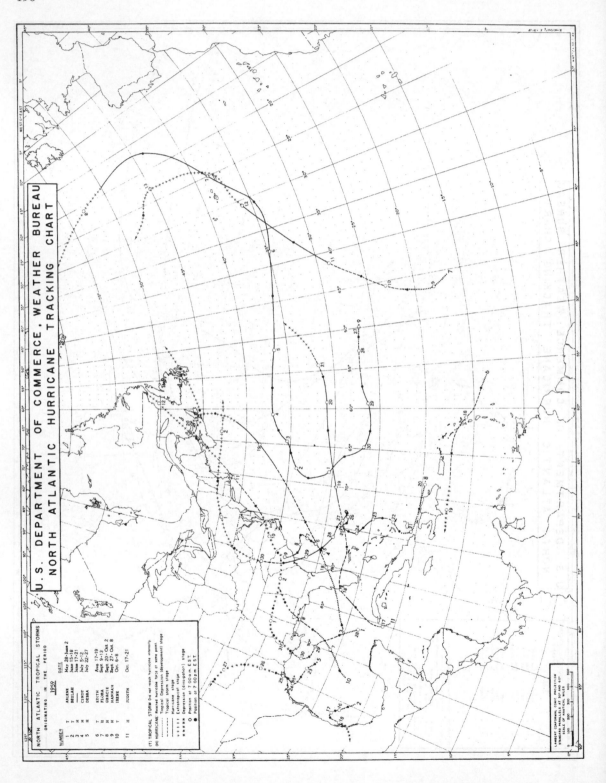

U.S. DEPARTMENT OF COMMERCE, WEATHER BUREAU
NORTH ATLANTIC HURRICANE TRACKING CHART

NORTH ATLANTIC TROPICAL STORMS
ORIGINATING IN THE PERIOD
1959

NUMBER	NAME	DATE
1	T	ARLENE	May 28-June 2
2	T	BEULAH	June 15-18
3	H		June 17-21
4	H	CINDY	July 5-12
5	H	DEBRA	July 22-27
6	T	EDITH	Aug. 17-19
7	H	FLORA	Sept. 9-13
8	H	GRACIE	Sept. 20-Oct. 2
9	H	HANNAH	Sept. 27-Oct. 8
10	T	IRENE	Oct. 6-8
11	H	JUDITH	Oct. 17-21

(T) TROPICAL STORM (Did not reach hurricane force at some point)
(H) HURRICANE Reached hurricane force at some point

· · · · · · · Tropical Depression (development) stage
- - - - - - - Tropical storm stage
+ + + + + + + Hurricane stage
x x x x x x x Extratropical stage
○ Depression (dissipation) stage
○ Position at 7:00 a.m. E.S.T.
● Position at 7:00 p.m. E.S.T.

LAMBERT CONFORMAL CONIC PROJECTION
STANDARD PARALLELS AT 30° AND 60°
SCALE OF NAUTICAL MILES
100 200 300 400 500

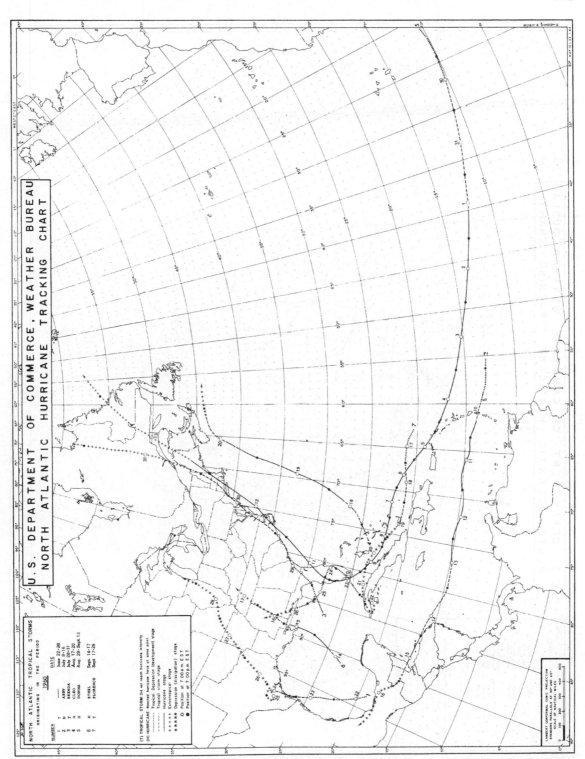

U.S. DEPARTMENT OF COMMERCE, WEATHER BUREAU
NORTH ATLANTIC HURRICANE TRACKING CHART

NORTH ATLANTIC TROPICAL STORMS
ORIGINATING IN THE PERIOD
1960

NUMBER		DATE
1	T	ABBY	June 22-26
2	H	BRENDA	July 9-16
3	T	CLEO	July 28-31
4	H	DONNA	Aug. 17-20
5	H		Aug. 29-Sept. 13
6	H	ETHEL	Sept. 14-17
7	T	FLORENCE	Sept. 17-26

(T) TROPICAL STORM Did not reach hurricane intensity.
(H) HURRICANE Reached hurricane force at some point.
.... Tropical Depression (development) stage
———— Tropical storm stage
———— Hurricane stage
++++ Extratropical stage
xxxx Depression (dissipation) stage
○ Position at 7:00 a.m. EST
● Position at 7:00 p.m. EST

192

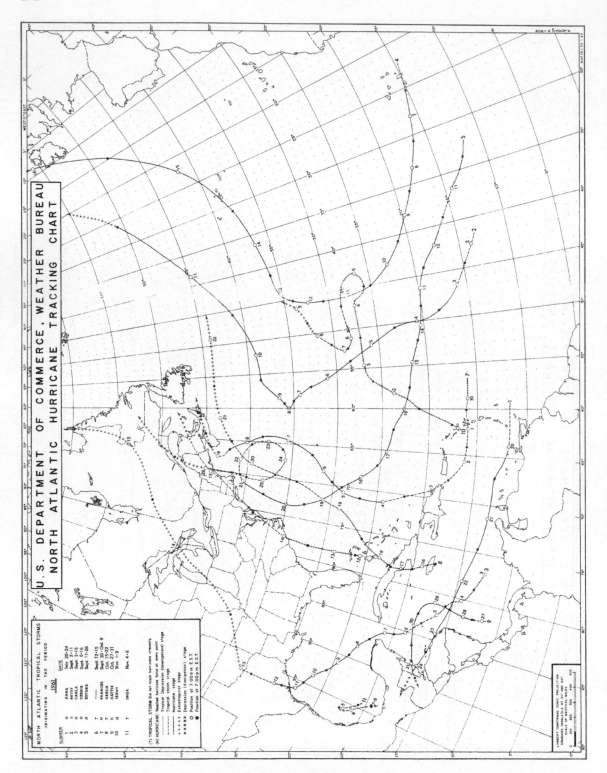

NORTH ATLANTIC TROPICAL STORMS
ORIGINATING IN THE PERIOD
1961

NUMBER		DATE
1	ANNA	July 20-24
2	BETSY	Sept. 2-11
3	CARLA	Sept. 3-15
4	DEBBIE	Sept. 6-16
5	ESTHER	Sept. 11-26
6	FRANCES	Sept. 12-15
7	GERDA	Sept. 20-Oct. 9
8	HATTIE	Oct. 15-22
9	INGA	Oct. 27-31
10	JENNY	Nov. 1-9
11	INGA	Nov. 4-8

(T) TROPICAL STORM Did not reach hurricane intensity

(H) HURRICANE Reached hurricane force at some point
.... Tropical Depression (development) stage
----------- Tropical storm stage
++++++++ Hurricane stage
××××××× Extratropical stage
***** Depression (dissipation) stage
○ Position at 7:00 a.m. E.S.T.
● Position at 7:00 p.m. E.S.T.

U.S. DEPARTMENT OF COMMERCE, WEATHER BUREAU
NORTH ATLANTIC HURRICANE TRACKING CHART

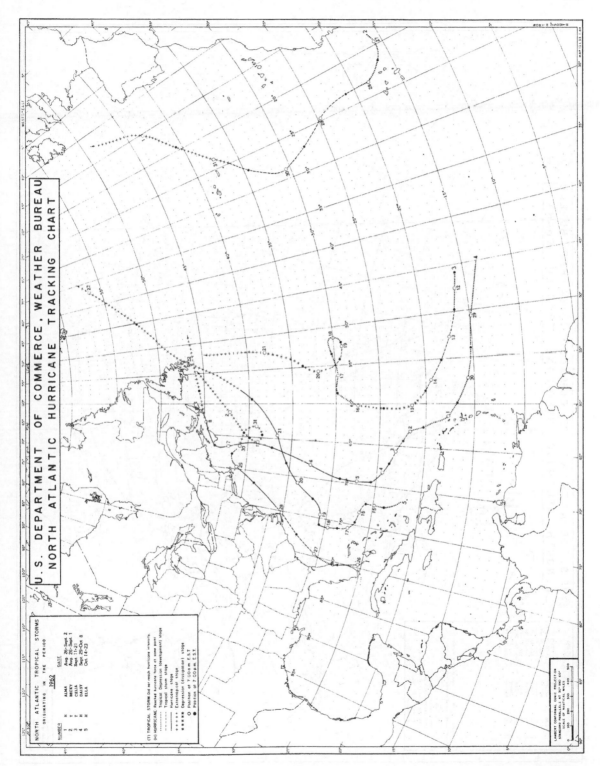

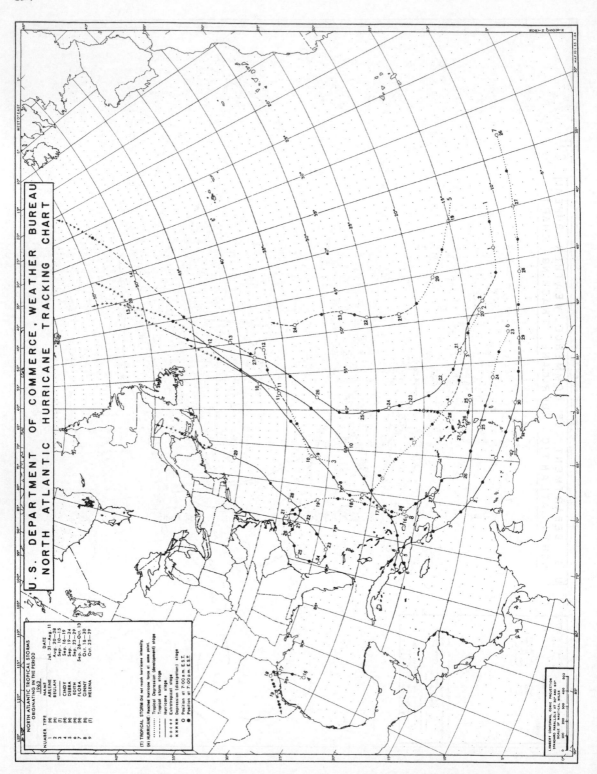

U.S. DEPARTMENT OF COMMERCE, WEATHER BUREAU
NORTH ATLANTIC HURRICANE TRACKING CHART

NORTH ATLANTIC TROPICAL STORMS
ORIGINATING IN THE PERIOD
1963

NUMBER	TYPE	NAME	DATE
1	(H)	ARLENE	Jul. 31–Aug. 11
2	(H)	BEULAH	Aug. 20–28
3	(T)	CINDY	Sep. 10–15
4	(H)	DEBRA	Sep. 19–24
5	(H)	EDITH	Sep. 23–29
6	(H)	FLORA	Sep. 26–Oct. 13
7	(H)	GINNY	Oct. 1–30
8	(T)	HELENA	Oct. 25–29

(T) TROPICAL STORM: Did not reach hurricane intensity.

(H) HURRICANE: Reached hurricane force at some point.

·········· Tropical (Depression (development) stage
+++++ Tropical storm stage
××××× Hurricane stage
━━━━━ Extratropical (dissipation) stage
○ Position at 7:00 a.m. E.S.T.
● Position at 7:00 p.m. E.S.T.

LAMBERT CONFORMAL CONIC PROJECTION
STANDARD PARALLELS AT 30° AND 60°
SCALE OF NAUTICAL MILES

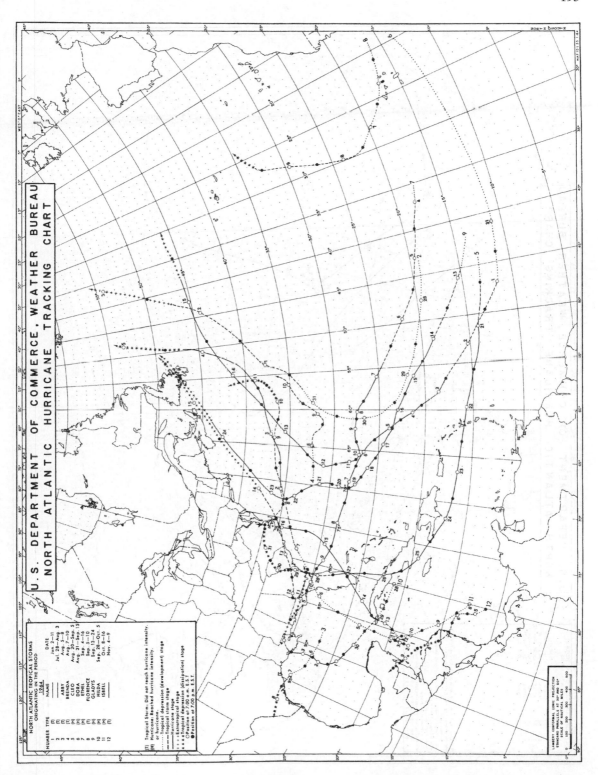

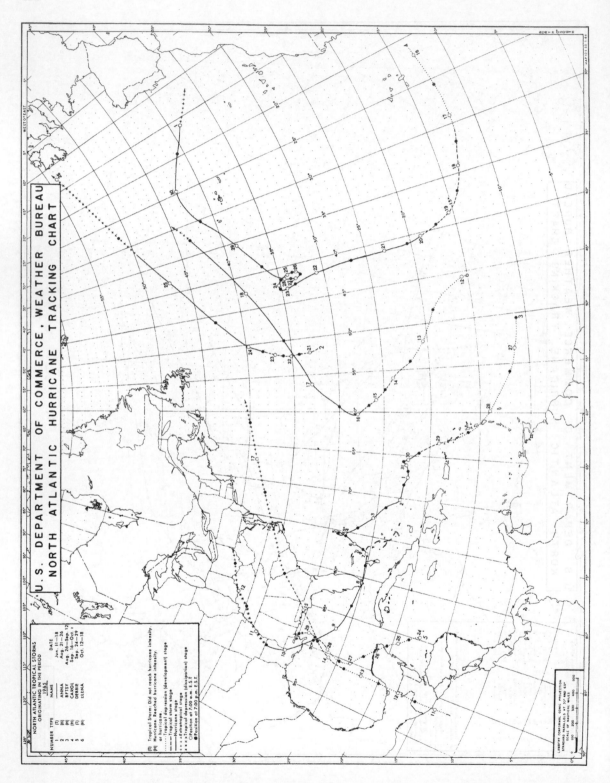

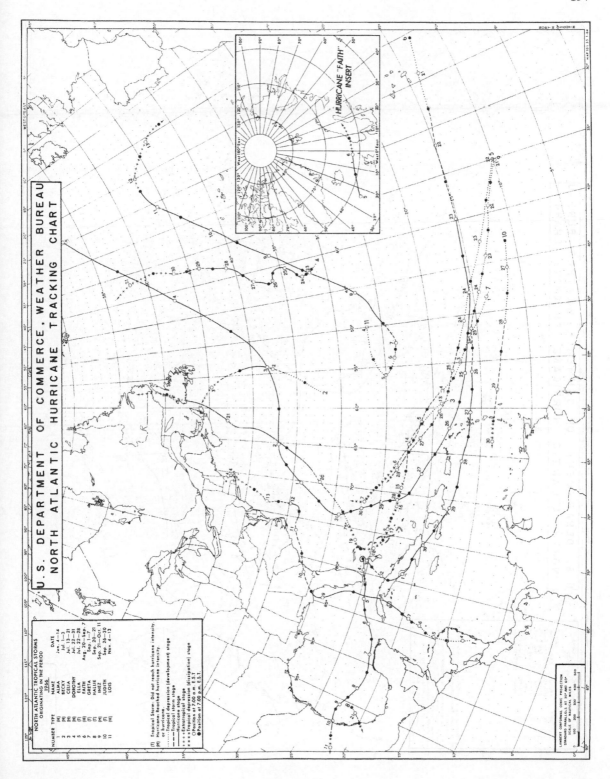

U.S. DEPARTMENT OF COMMERCE, WEATHER BUREAU
NORTH ATLANTIC HURRICANE TRACKING CHART

HURRICANE "FAITH" INSERT

NORTH ATLANTIC TROPICAL STORMS
ORIGINATING IN THE PERIOD
1966

NUMBER	TYPE	NAME	DATE
1	(H)	ALMA	Jun. 4—14
2	(H)	BECKY	Jul. 1—3
3	(H)	CELIA	Jul. 13—21
4	(H)	DOROTHY	Jul. 22—31
5	(H)	ELLA	Jul. 29—28
6	(H)	FAITH	Aug. 20—Sep. 7
7	(H)	GRETA	Sep. 1—7
8	(T)	HALLIE	Sep. 20—21
9	(T)	INEZ	Sep. 21—Oct. 11
10	(T)	JUDITH	Sep. 26—30
11	(H)	LOIS	Nov. 4—13

(T) Tropical Storm. Did not reach hurricane intensity.
(H) Hurricane. Reached hurricane intensity.

● ● ● ● ● at hurricane intensity.
—— —— Tropical depression (development) stage
———— Tropical storm stage
———— Hurricane stage
· · · · · Extratropical stage
× × × × × Tropical depression (dissipation) stage
○ Position at 7:00 a.m. E.S.T.
● Position at 7:00 p.m. E.S.T.

LAMBERT CONFORMAL CONIC PROJECTION
STANDARD PARALLELS AT 30° AND 60°
SCALE OF NAUTICAL MILES

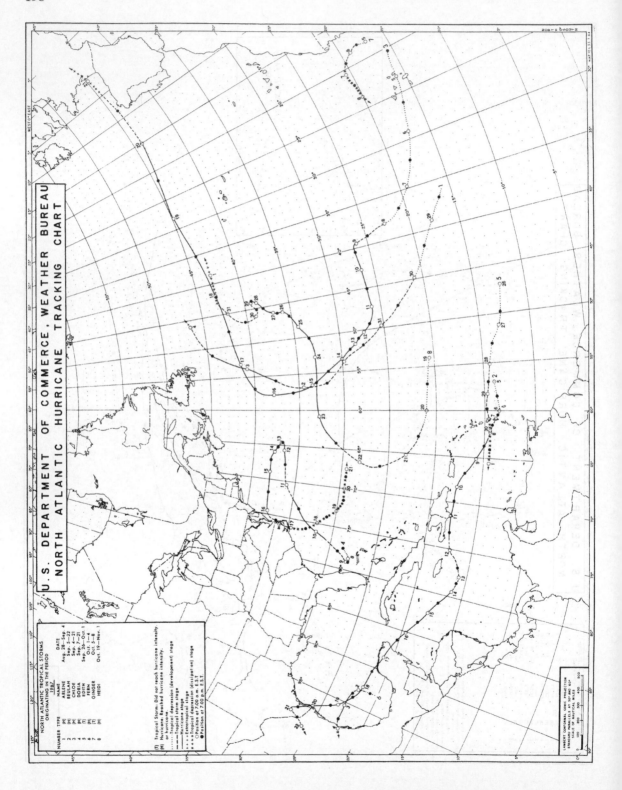

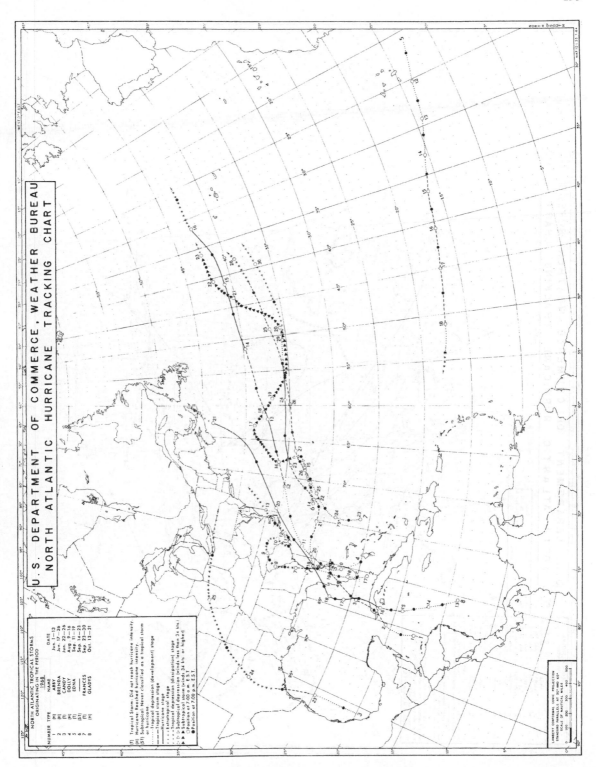

U.S. DEPARTMENT OF COMMERCE, WEATHER BUREAU
NORTH ATLANTIC HURRICANE TRACKING CHART

NORTH ATLANTIC TROPICAL STORMS
ORIGINATING IN THE PERIOD
1968

NUMBER	TYPE	NAME	DATE
1	(H)	ABBY	Jun. 1—13
2	(H)	BRENDA	Jun. 17—26
3	(H)	CANDY	Jun. 22—26
4	(H)	DOLLY	Aug. 9—16
5	(H)	EDNA	Sep. 11—19
6	(ST)		Sep. 14—23
7	(T)	FRANCES	Sep. 21—30
8	(H)	GLADYS	Oct. 13—21

(T) Tropical Storm. Did not reach hurricane intensity
(H) Hurricane. Reached hurricane intensity.
(ST) Subtropical. Never classified as a tropical storm
or hurricane.

········ Tropical depression (development) stage
————— Tropical storm stage
————— Hurricane stage
· · · · Extratropical stage
▲▲▲▲ Tropical depression (dissipation) stage
▷▷▷▷ Subtropical depression (winds less than 34 kts.)
▲▲▲▲ Subtropical storm (winds 34 kts. or higher)
○ Position at 7:00 a.m. E.S.T.
● Position at 7:00 p.m. E.S.T.

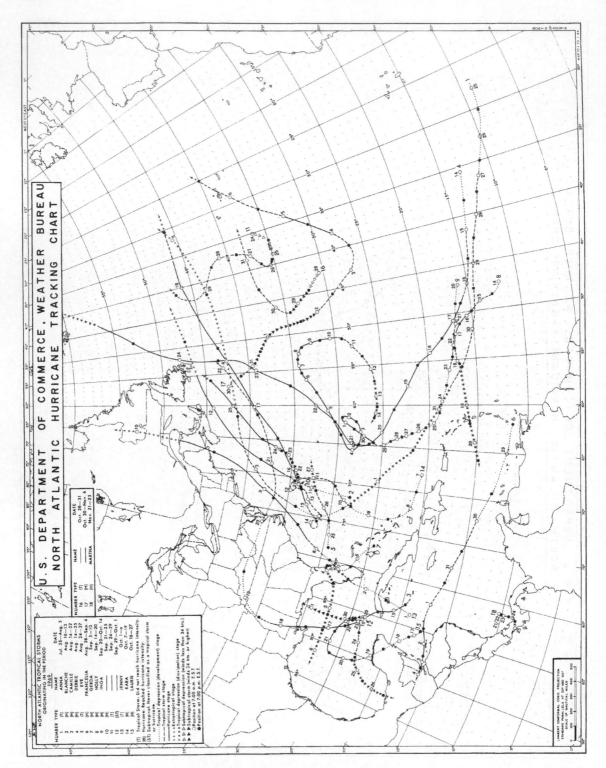

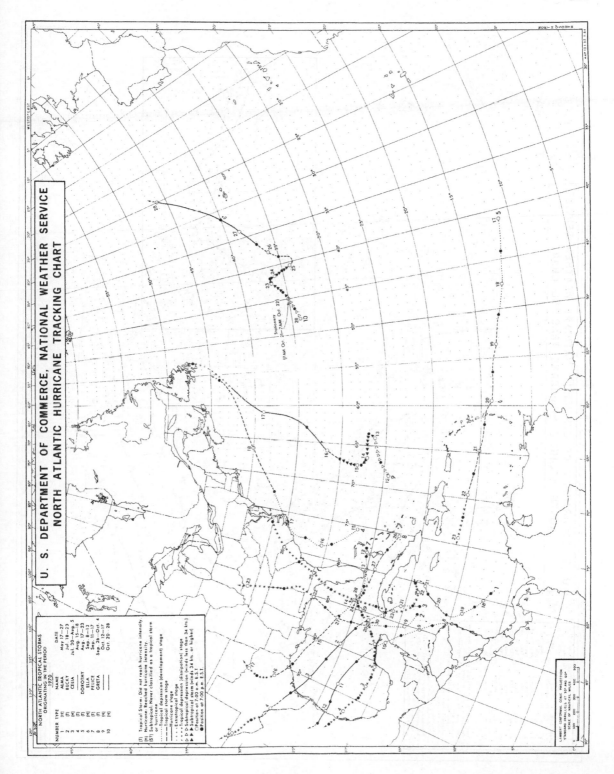

U. S. DEPARTMENT OF COMMERCE, NATIONAL WEATHER SERVICE
NORTH ATLANTIC HURRICANE TRACKING CHART

NORTH ATLANTIC TROPICAL STORMS
ORIGINATING IN THE PERIOD
1970

NUMBER	TYPE	NAME	DATE
1	(T)	ALMA	May 17–27
2	(H)	BECKY	Jul 18–23
3	(T)	CELIA	Jul 30–Aug. 5
4	(H)		Aug. 15–18
5	(H)		Aug. 17–23
6	(T)	DOROTHY	Sep. 8–13
7	(H)	ELLA	Sep. 11–17
8	(T)	FELICE	Sep. 26–Oct. 4
9	(H)	GRETA	Oct. 2–17
10	(H)		Oct. 20–28

(T) Tropical Storm. Did not reach hurricane intensity.
(H) Hurricane. Reached hurricane intensity.
(ST) Subtropical. Never classified as a tropical storm or hurricane.

- - - - - - Tropical depression (development) stage
· · · · · · · Tropical storm stage
————— Hurricane stage
· · · · · · · · Extratropical stage
· · · · · · · · Tropical (or subtropical) dissipating stage
◁ ▷ ▽ Subtropical depression (winds less than 34 kn.)
▲ ▶ ▼ Subtropical storm (winds 34 kn. or higher)
◯ Position at 7:00 a.m. E.S.T.
● Position at 7:00 p.m. E.S.T.

LAMBERT CONFORMAL CONIC PROJECTION
STANDARD PARALLELS AT 30° AND 60°
SCALE OF NAUTICAL MILES

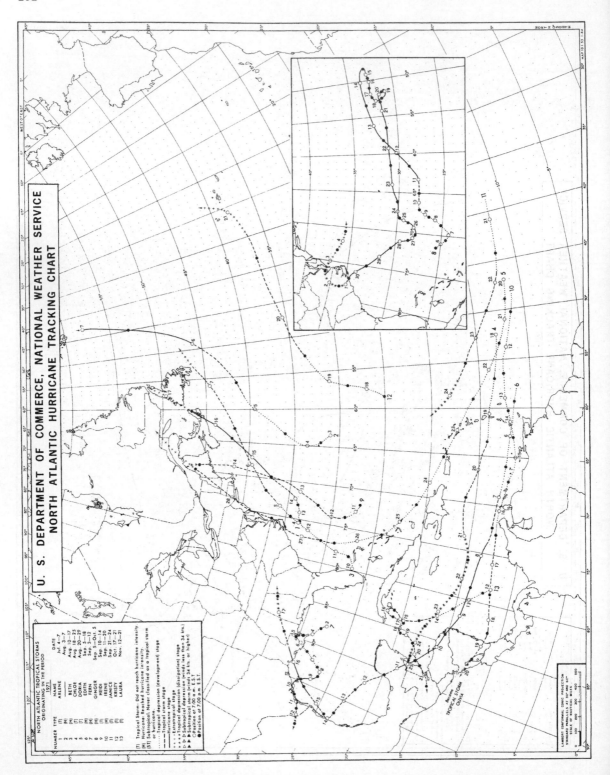

U. S. DEPARTMENT OF COMMERCE, NATIONAL WEATHER SERVICE
NORTH ATLANTIC HURRICANE TRACKING CHART

NORTH ATLANTIC TROPICAL STORMS
ORIGINATING IN THE PERIOD
1971

NUMBER	TYPE	NAME	DATE
1	[T]	ARLENE	Jul. 4—7
2	[H]	BETH	Aug. 3—7
3	[H]	CHLOE	Aug. 10—17
4	[T]	DORIA	Aug. 20—29
5	[H]	EDITH	Sep. 5—18
6	[H]	FERN	Sep. 3—12
7	[H]	GINGER	Sep. 10—Oct. 5
8	[H]	HEIDI	Sep. 10—14
9	[H]	IRENE	Sep. 11—20
10	[T]	JANICE	Sep. 21—24
11	[T]	KRISTY	Oct. 17—21
12	[T]	LAURA	Nov. 12—21
13	[T]		

[T] Tropical Storm: Did not reach hurricane intensity.
[H] Hurricane: Reached hurricane intensity.
[ST] Subtropical: Never classified as a tropical storm or hurricane.

——— Tropical depression (development) stage
.... Tropical storm stage
——— Hurricane stage
+++++ Extratropical stage
▷ ▷ ▷ Tropical depression (dissipation) stage
△ ▲ Subtropical depression (winds less than 34 kts.)
▲ ▲ Subtropical storm (winds 34 kts. or higher)
○ Position at 7:00 a.m. E.S.T.
● Position at 7:00 p.m. E.S.T.

LAMBERT CONFORMAL CONIC PROJECTION
STANDARD PARALLELS AT 30° AND 60°
SCALE OF NAUTICAL MILES
0 100 200 300 400 500

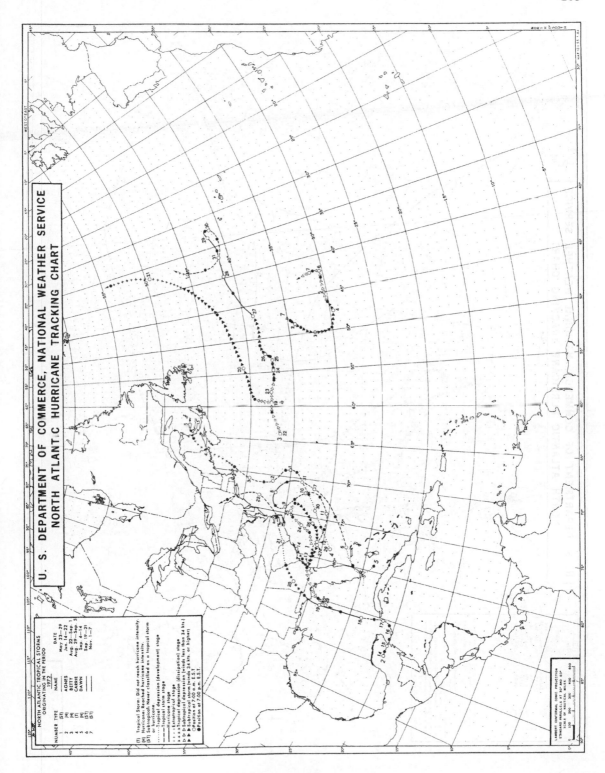

U. S. DEPARTMENT OF COMMERCE, NATIONAL WEATHER SERVICE
NORTH ATLANTIC HURRICANE TRACKING CHART

NORTH ATLANTIC TROPICAL STORMS
ORIGINATING IN THE PERIOD
1972

NUMBER	TYPE	NAME	DATE
2	(ST)	AGNES	May 23—29
3	(H)	BETTY	Jun. 14—22
4	(H)	CARRIE	Aug. 29—Sep. 5
5	(T)	DAWN	Aug. 22—Sep. 1
6	(ST)		Sep. 4—14
7	(ST)		Sep. 19—21
			Nov. 1—7

(T) Tropical Storm: Did not reach hurricane intensity.
(H) Hurricane: Reached hurricane intensity.
(ST) Subtropical: Never classified as a tropical storm
or a hurricane.
········· Tropical depression (development) stage
········· Tropical depression stage
------- Extratropical stage
▲▲▲ Hurricane stage
⊳⊳⊳ Subtropical depression (dissipation) stage
△△△ Tropical depression (winds less than 34 kts.)
▲▲▲ Subtropical storm (winds 34 kts. or higher)
○ Position at 7:00 a.m. E.S.T.
● Position at 7:00 p.m. E.S.T.

LAMBERT CONFORMAL CONIC PROJECTION
STANDARD PARALLELS AT 30° AND 60°
SCALE OF NAUTICAL MILES
100 200 300 400 500

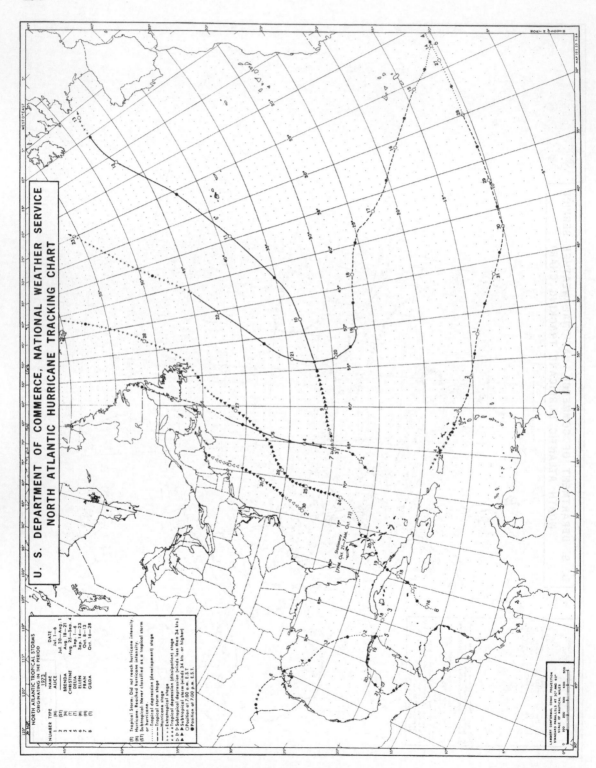

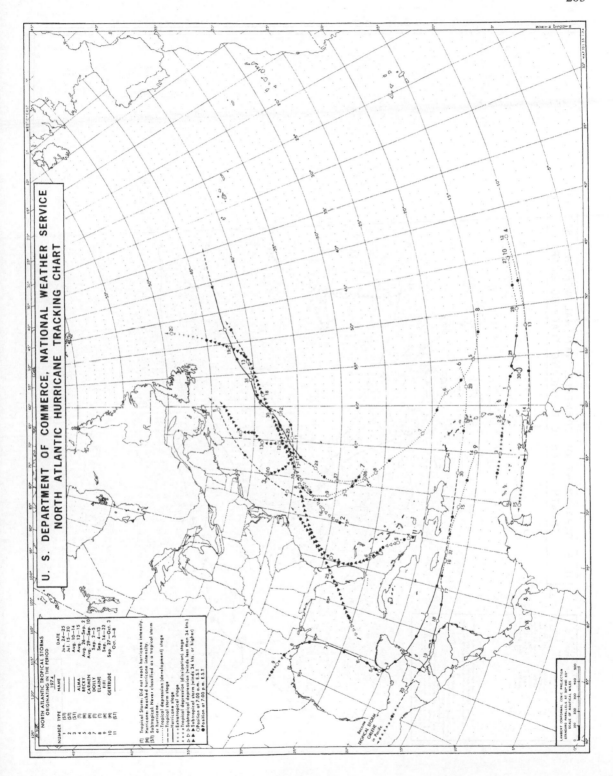

U. S. DEPARTMENT OF COMMERCE, NATIONAL WEATHER SERVICE
NORTH ATLANTIC HURRICANE TRACKING CHART

NORTH ATLANTIC TROPICAL STORMS
ORIGINATING IN THE PERIOD
1974

NUMBER	TYPE	NAME	DATE
1	(ST)		Jun. 24–25
2	(ST)		Jul. 15–20
3	(ST)		Aug. 10–14
4	(T)	ALMA	Aug. 12–15
5	(H)	BECKY	Aug. 26–Sep. 2
6	(H)	CARMEN	Aug. 29–Sep. 10
7	(T)	DOLLY	Sep. 2–5
8	(T)	ELAINE	Sep. 4–13
9	(H)	FIFI	Sep. 16–22
10	(H)		Sep. 27–Oct. 3
11	(ST)	GERTRUDE	Oct. 1–8

(T) Tropical Storm: Did not reach hurricane intensity.
(H) Hurricane: Reached hurricane intensity
(ST) Subtropical: Never classified as a tropical storm
     or hurricane
.... Tropical depression (development) stage
——— Tropical storm stage
+ + + + Extratropical stage
• • • • Tropical depression (dissipation) stage
△ ▷ ▷ Subtropical depression (winds less than 34 kts.)
△ ▲ ▲ Subtropical storm (winds 34 kts. or higher)
○ Position at 7:00 a.m. E.S.T
● Position at 7:00 p.m. E.S.T

LAMBERT CONFORMAL CONIC PROJECTION
STANDARD PARALLELS AT 30° AND 60°
SCALE OF NAUTICAL MILES

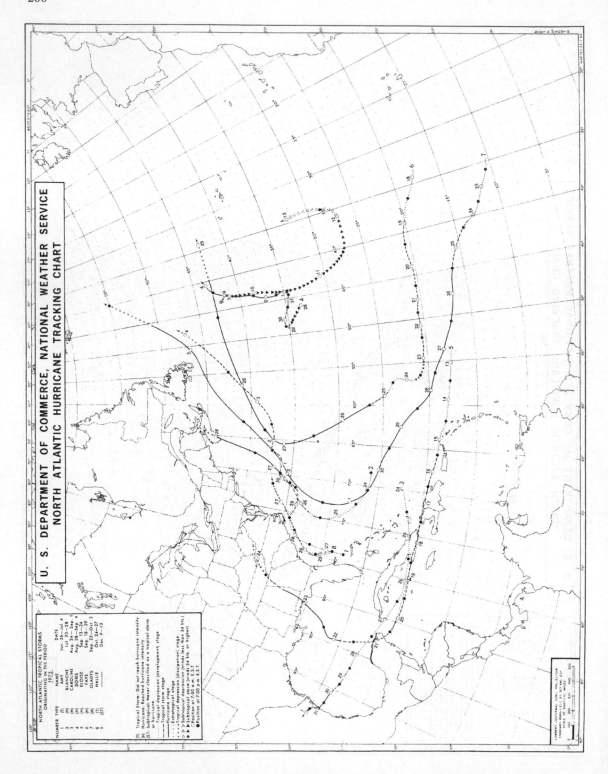

U. S. DEPARTMENT OF COMMERCE, NATIONAL WEATHER SERVICE
NORTH ATLANTIC HURRICANE TRACKING CHART

NORTH ATLANTIC TROPICAL STORMS
ORIGINATING IN THE PERIOD 1975

NUMBER TYPE	NAME	DATE
1	(T)	AMY	Jun. 26—Jul. 4
2	(T)	BLANCHE	Jul. 23—28
3	(H)	CAROLINE	Aug. 24— Sep. 1
4	(H)	DORIS	Aug. 28—Sep. 4
5	(H)	ELOISE	Sep. 13—24
6	(T)	FAYE	Sep. 18—24
7	(H)	GLADYS	Sep. 22—Oct. 3
8	(H)	HALLIE	Oct. 24—27
9	(ST)		Dec. 9—13

(T) Tropical Storm Did not reach hurricane intensity
(H) Hurricane Reached hurricane intensity
(ST) Subtropical Never classified as a tropical storm
        or hurricane
⋯⋯⋯ Tropical depression (development) stage
—— Tropical storm stage
———— Hurricane stage
−−−− Extratropical stage
∙∙∙∙∙ Tropical depression (dissipation) stage
△ ▷ ◁ Subtropical depression stage
▲ ▶ Subtropical storm (winds less than 34 kts.)
▲ ▲ Subtropical depression (winds 34 kts. or higher)
○ Position at 7:00 a.m. E.S.T.
● Position at 7:00 p.m. E.S.T.

LAMBERT CONFORMAL CONIC PROJECTION
STANDARD PARALLELS AT 30° AND 60°
SCALE OF NAUTICAL MILES

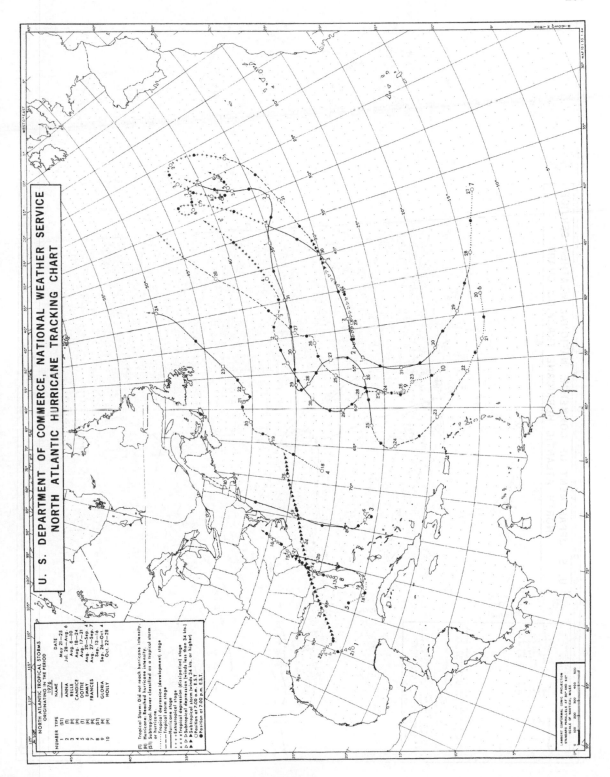

U. S. DEPARTMENT OF COMMERCE, NATIONAL WEATHER SERVICE
NORTH ATLANTIC HURRICANE TRACKING CHART

NORTH ATLANTIC TROPICAL STORMS
ORIGINATING IN THE PERIOD
1976

NUMBER	TYPE	NAME	DATE
1	[ST]	ANNA	May 21–25
2	[T]		Jul. 28–Aug. 6
3	[H]	BELLE	Aug. 6–10
4	[H]	CANDICE	Aug. 18–24
5	[T]	DOTTIE	Aug. 18–22
6	[H]	EMMY	Aug. 20–Sep. 4
7	[H]	FRANCES	Aug. 27–Sep. 7
8	[ST]		Sep. 13–16
9	[H]	GLORIA	Sep. 26–Oct. 4
10	[H]	HOLLY	Oct. 22–28

[T] Tropical Storm. Did not reach hurricane intensity.
[H] Hurricane. Reached hurricane intensity.
[ST] Subtropical. Never classified as a tropical storm or hurricane.

········ Tropical depression (development) stage
──── Tropical storm stage
──── Hurricane stage
++++ Extratropical stage
••••• Tropical depression (dissipation) stage
△ ○ Subtropical depression (winds less than 34 kts.)
△ ▲ Subtropical storm (winds 34 kts. or higher)
○ Position at 7:00 a.m. E.S.T.
● Position at 7:00 p.m. E.S.T.

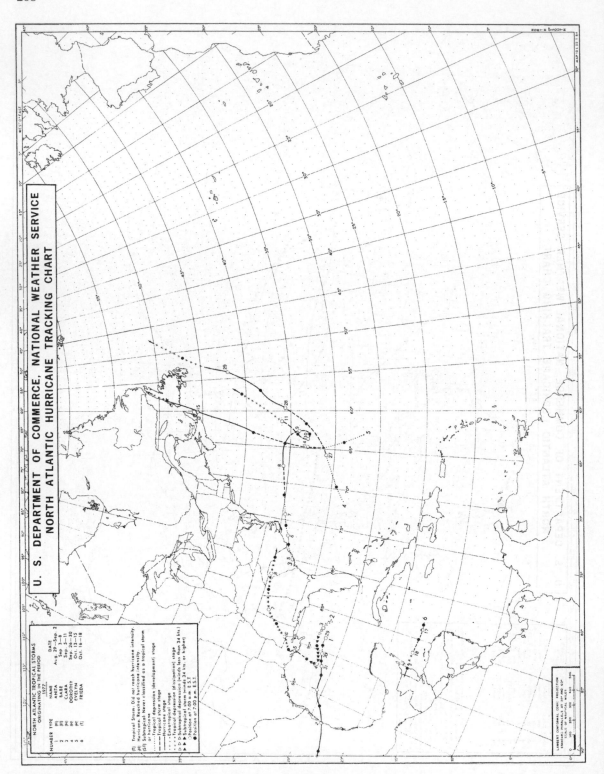

U. S. DEPARTMENT OF COMMERCE, NATIONAL WEATHER SERVICE
NORTH ATLANTIC HURRICANE TRACKING CHART

NORTH ATLANTIC TROPICAL STORMS
1977
ORIGINATING IN THE PERIOD

NUMBER	TYPE	NAME	DATE
1	(H)	ANITA	Aug. 29–Sep. 2
2	(H)	BABE	Sep. 3–8
3	(H)	CLARA	Sep. 5–11
4	(H)	DOROTHY	Sep. 26–30
5	(H)	EVELYN	Oct. 13–15
6	(F)	FRIEDA	Oct. 16–18

(F) Tropical Storm Did not reach hurricane intensity.
(H) Hurricane Reached hurricane intensity.
(ST) Subtropical Never classified as a tropical storm or hurricane.

LAMBERT CONFORMAL CONIC PROJECTION
STANDARD PARALLELS AT 30° AND 60°
SCALE OF NAUTICAL MILES

209

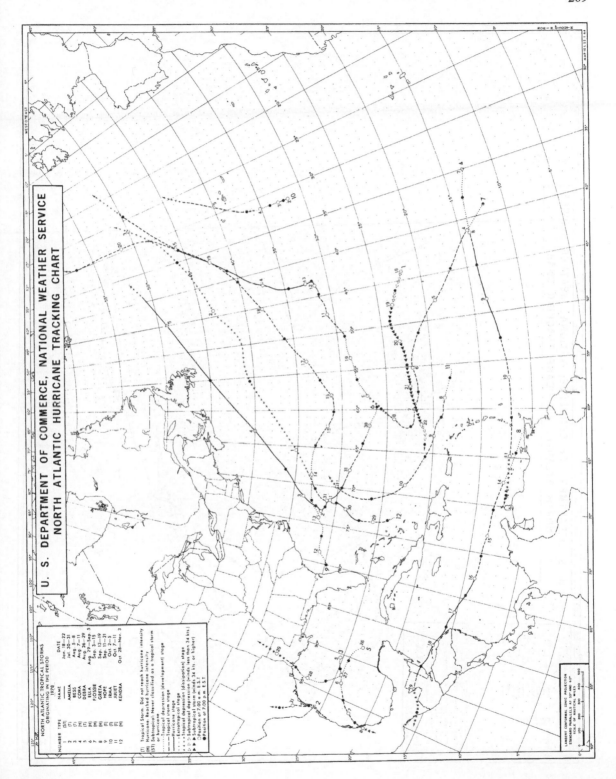

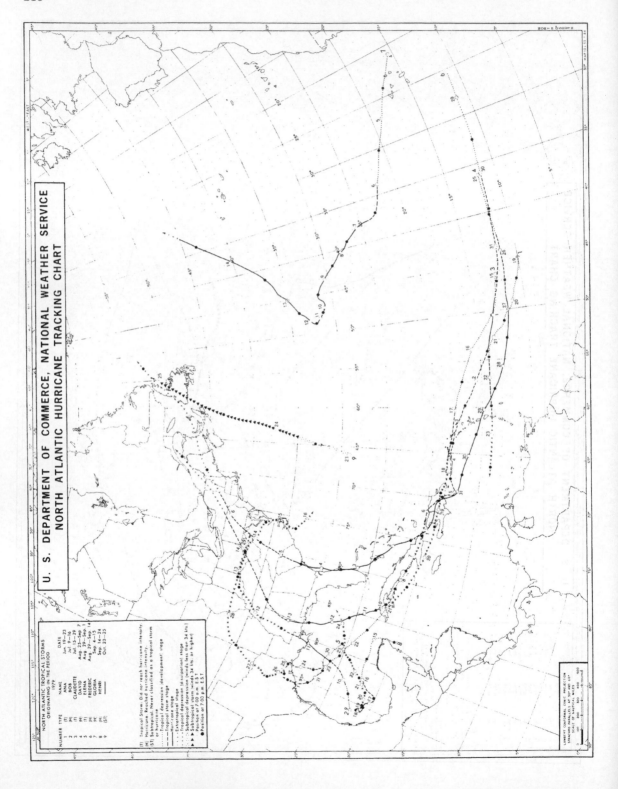

U. S. DEPARTMENT OF COMMERCE, NATIONAL WEATHER SERVICE
NORTH ATLANTIC HURRICANE TRACKING CHART

NORTH ATLANTIC TROPICAL STORMS
ORIGINATING IN THE PERIOD

1979

NUMBER	TYPE	NAME	DATE
1	(T)	ANA	Jun 19—23
2	(H)	BOB	Jul 9—16
3	(T)	CLAUDETTE	Jul 15—29
4	(H)	DAVID	Aug 25—Sep 7
5	(H)	ELENA	Aug 29—Sep 1
6	(H)	FREDERIC	Aug 29—Sep 14
7	(H)	GLORIA	Sep 14—15
8	(H)	HENRI	Oct. 23—25
9	(ST)		

(T) Tropical Storm: Did not reach hurricane intensity.
(H) Hurricane: Reached hurricane intensity.
(ST) Subtropical: Never classified as a tropical storm or hurricane.

---- Tropical depression development stage
———— Tropical storm stage
———— Hurricane stage
------ Extratropical stage
········ Subtropical depression (dissipation) stage
········ Subtropical depression (winds less than 34 kts)
▲▲▲▲ Subtropical storm (winds 34 kts or higher)
△ Position at 7:00 a.m. E.S.T
● Position at 7:00 p.m. E.S.T

LAMBERT CONFORMAL CONIC PROJECTION
STANDARD PARALLELS AT 30° AND 60°
SCALE OF NAUTICAL MILES

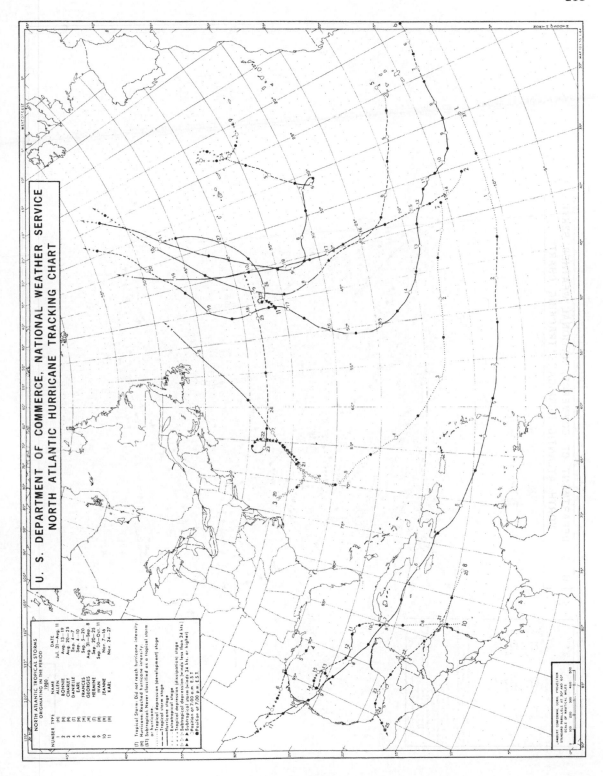

U. S. DEPARTMENT OF COMMERCE, NATIONAL WEATHER SERVICE
NORTH ATLANTIC HURRICANE TRACKING CHART

NORTH ATLANTIC TROPICAL STORMS
ORIGINATING IN THE PERIOD
1980.

NUMBER	TYPE	NAME	DATE
1	(H)	ALLEN	Jul. 31–Aug. 11
2	(H)	BONNIE	Aug. 13–19
3	(H)	CHARLEY	Aug. 20–25
4	(H)	DANIELLE	Sep. 4–7
5	(T)	EARL	Sep. 5–20
6	(H)	FRANCES	Sep. 5–20
7	(H)	GEORGES	Aug. 31–Sep. 8
8	(T)	HERMINE	Sep. 20–25
9	(H)	IVAN	Sep. 30–Oct. 11
10	(T)	JEANNE	Nov. 7–16
11	(H)	KARL	Nov. 24–27

(T) Tropical Storm: Did not reach hurricane intensity
(H) Hurricane: Reached hurricane intensity
(ST) Subtropical: Never classified as a tropical storm or hurricane

.... Tropical depression (development) stage
———— Tropical storm stage
———— Hurricane stage
━━━━ Extratropical stage
∙∙∙∙∙∙ Tropical depression (dissipation) stage
○○○○ Subtropical depression (winds less than 34 kts.)
▲▲▲▲ Subtropical storm (winds 34 kt. or higher)
△ Position at 7:00 a.m. E.S.T
● Position at 7:00 p.m. E.S.T

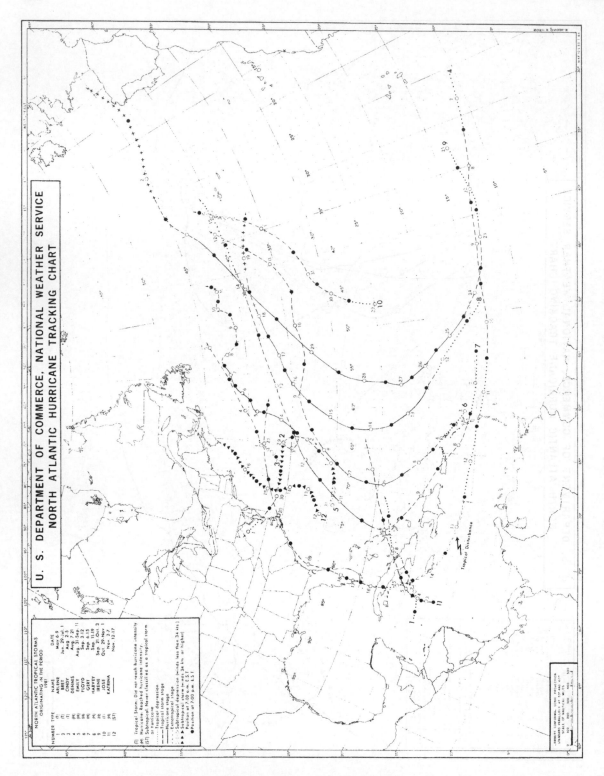

U. S. DEPARTMENT OF COMMERCE, NATIONAL WEATHER SERVICE
NORTH ATLANTIC HURRICANE TRACKING CHART

NORTH ATLANTIC TROPICAL STORMS
ORIGINATING IN THE PERIOD
1981

NUMBER	TYPE	NAME	DATE
1	(T)	ARLENE	May 6-9
2	(T)	BRET	Jun. 29-Jul. 1
3	(T)	CINDY	Aug. 2-5
4	(H)	DENNIS	Aug. 7-21
5	(H)	EMILY	Aug. 31-Sep. 11
6	(H)	FLOYD	Sep. 3-12
7	(H)	GERT	Sep. 6-15
8	(H)	HARVEY	Sep. 11-19
9	(H)	IRENE	Sep. 21-Oct. 3
10	(H)	JOSE	Oct. 29-Nov. 1
11	(H)	KATRINA	Nov. 2-7
12	(ST)		Nov. 12-17

(T) Tropical Storm. Did not reach hurricane intensity
(H) Hurricane. Reached hurricane intensity
(ST) Subtropical. Never classified as a tropical storm

Tropical depression
Tropical storm stage
Hurricane stage
Extratropical stage
Subtropical depression (winds less than 34 kts)
Subtropical storm (winds 34 kts. or higher)
Position at 7:00 a.m. E.S.T
Position at 7:00 p.m. E.S.T

Tropical Disturbance

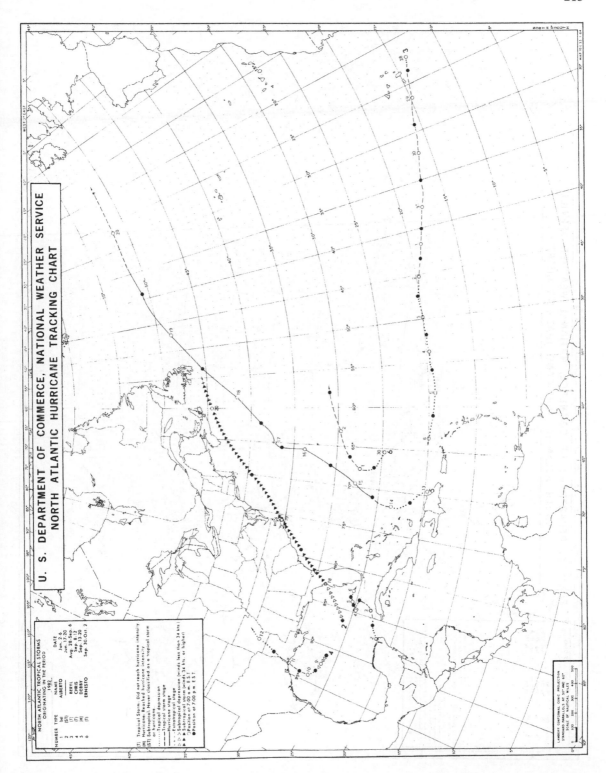

U. S. DEPARTMENT OF COMMERCE, NATIONAL WEATHER SERVICE
NORTH ATLANTIC HURRICANE TRACKING CHART

NORTH ATLANTIC TROPICAL STORMS
ORIGINATING IN THE PERIOD 1982

NUMBER	TYPE	NAME	DATE
1	(H)	ALBERTO	Jun. 2-6
2	(ST)	BERYL	Jun. 17-20
3	(ST)	CHRIS	Aug. 28-Sep. 6
4	(H)	DEBBY	Sep. 8-12
5	(H)	ERNESTO	Sep. 13-20
6	(T)		Sep. 30-Oct. 2

(T) Tropical Storm: Did not reach hurricane intensity
(H) Hurricane: Reached hurricane intensity
(ST) Subtropical: Never classified as a tropical storm or hurricane

.... Tropical depression
———— Tropical storm stage
———— Hurricane stage
—·—·— Extratropical stage
▷ ◁ Subtropical depression (winds less than 34 kts.)
▲ ▲ Subtropical storm (winds 34 kts. or higher)
◯ Position at 7:00 a.m. EST
● Position at 7:00 p.m. EST

LAMBERT CONFORMAL CONIC PROJECTION
STANDARD PARALLELS AT 30° AND 60°
SCALE OF NAUTICAL MILES

214

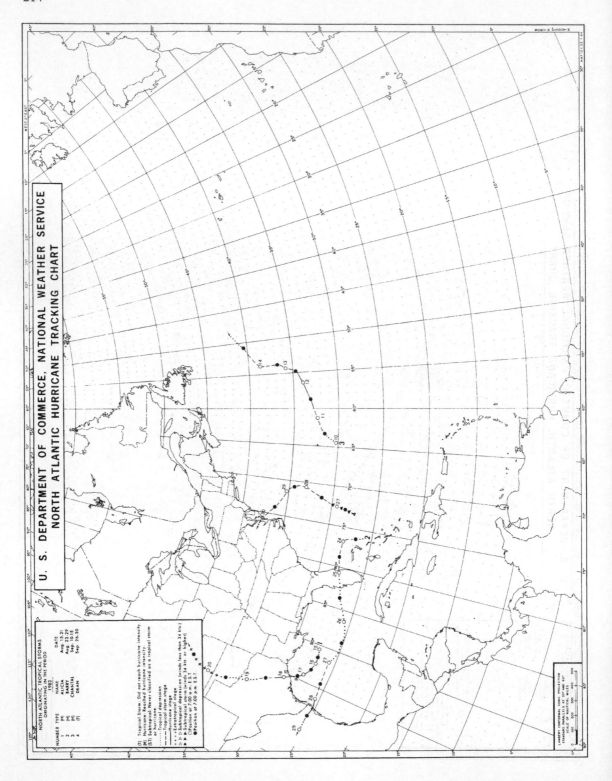

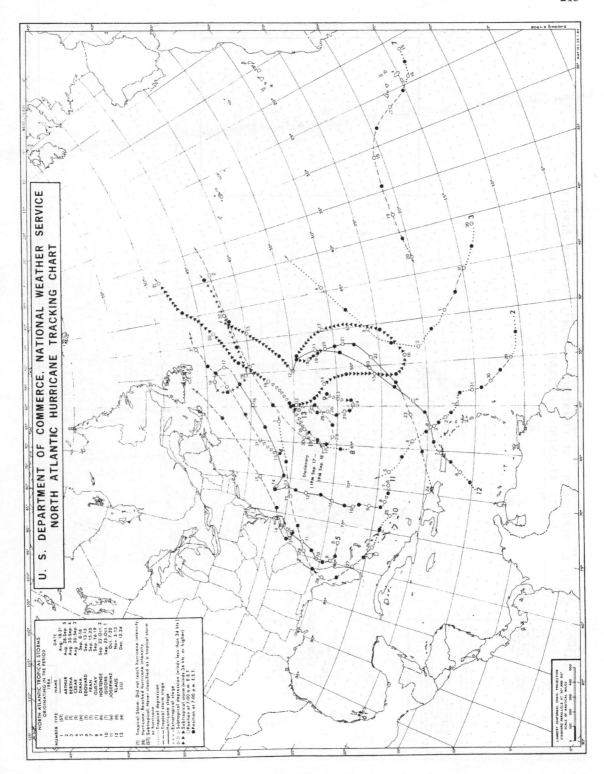

U. S. DEPARTMENT OF COMMERCE, NATIONAL WEATHER SERVICE
NORTH ATLANTIC HURRICANE TRACKING CHART

216

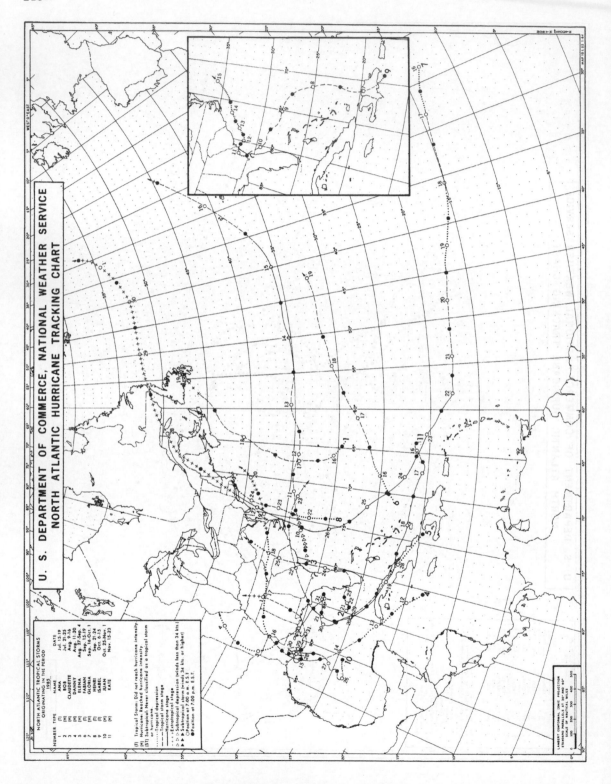

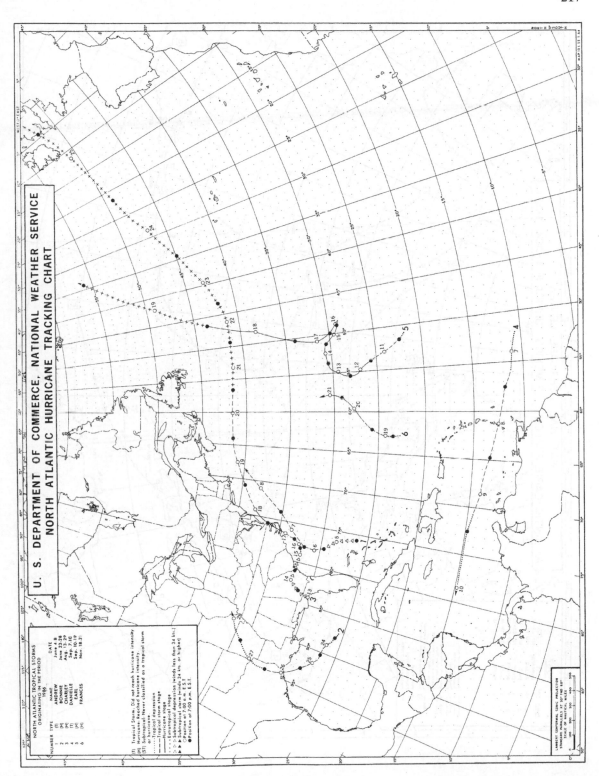

U. S. DEPARTMENT OF COMMERCE, NATIONAL WEATHER SERVICE
NORTH ATLANTIC HURRICANE TRACKING CHART

218

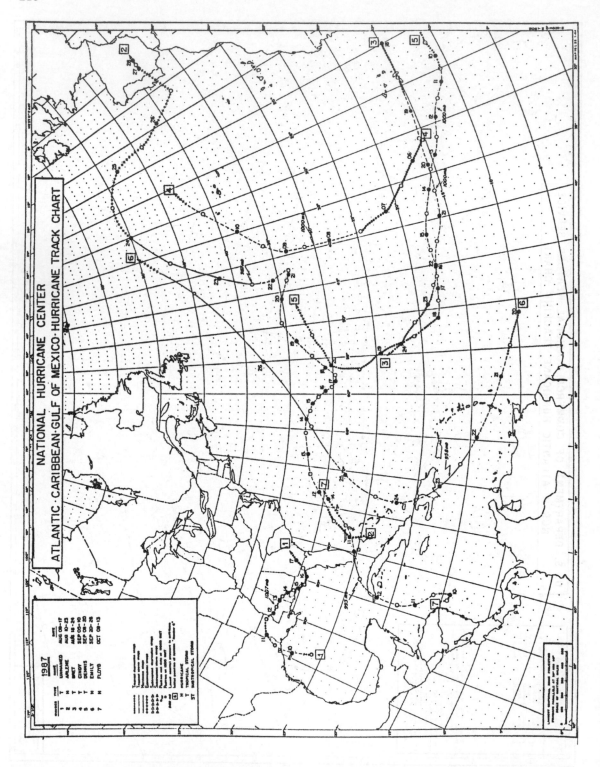

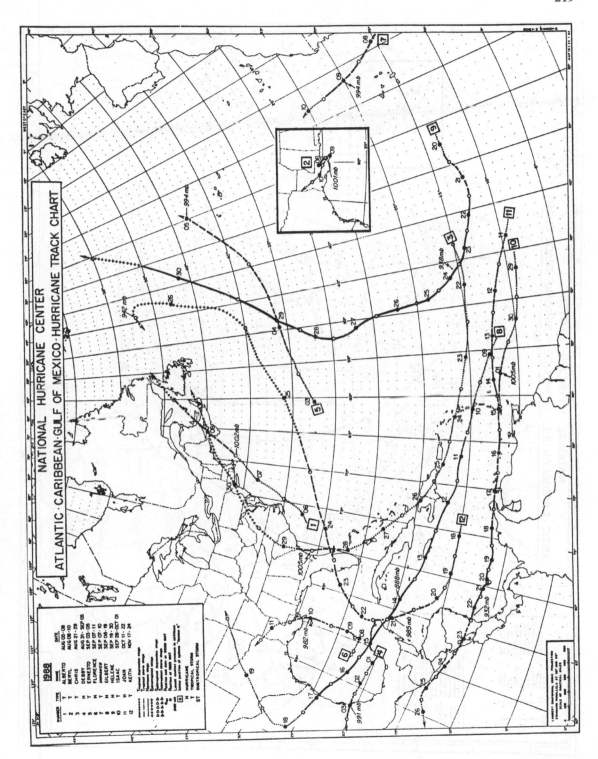

NATIONAL HURRICANE CENTER
ATLANTIC · CARIBBEAN · GULF OF MEXICO · HURRICANE TRACK CHART

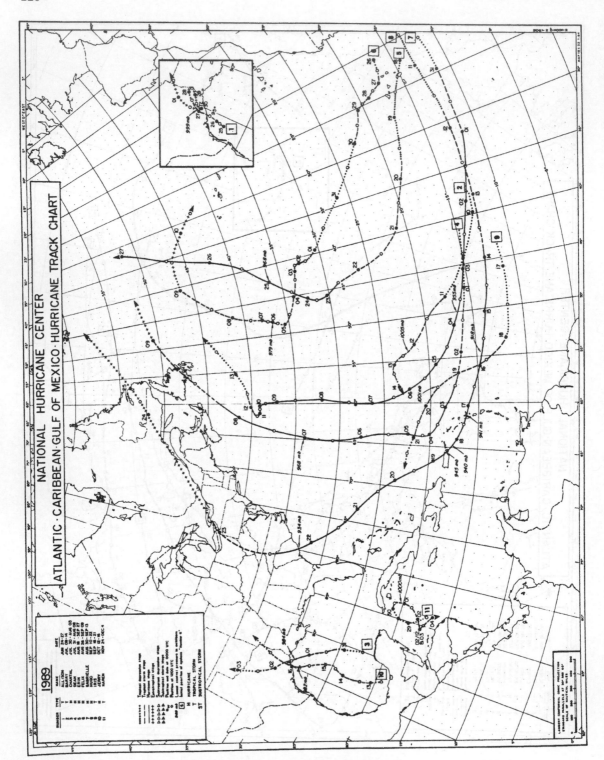

Appendix B

ATLANTIC TROPICAL CYCLONE TRACK MAP

Source: US Government Publication

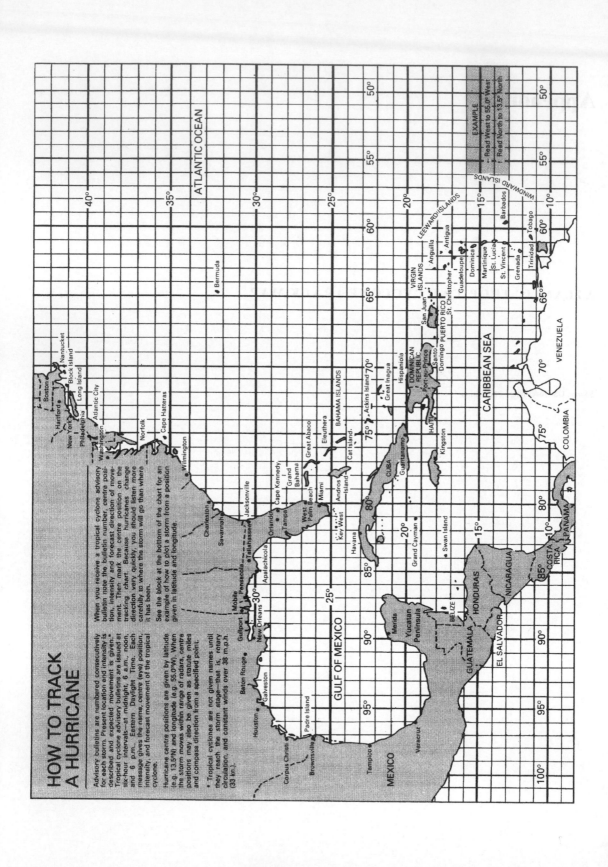

HOW TO TRACK A HURRICANE

Advisory bulletins are numbered consecutively for each storm. Present location and intensity is described and expected movement is given.* Tropical cyclone advisory bulletins are issued at six-hour intervals—at midnight, 6 a.m., noon, and 6 p.m., Eastern Daylight Time. Each message gives the name, centre (eye) position, intensity, and forecast movement of the tropical cyclone.

Hurricane centre positions are given by latitude (e.g. 13.5°N) and longitude (e.g. 55.0°W). When the storm moves within range of radars, centre positions may also be given as statute miles and compass direction from a specified point.

When you receive a tropical cyclone advisory bulletin note the bulletin number, centre position, intensity and forecast direction of movement. Then mark the centre position on the tracking chart. Because hurricanes change direction very quickly, you should listen more carefully to where the storm will go than where it has been.

See the block at the bottom of the chart for an example of how to plot a storm from a position given in latitude and longitude.

* Tropical cyclones are not given names until they reach the storm stage—that is, rotary circulation and constant winds over 38 m.p.h. (33 kn.).

EXAMPLE
—Read West to 55.0° West
—Read North to 13.5° North

ATLANTIC OCEAN

GULF OF MEXICO

CARIBBEAN SEA

WINDWARD ISLANDS

LEEWARD ISLANDS

VIRGIN ISLANDS

BAHAMA ISLANDS

References

Anthes, R.A. (1982) *Tropical Cyclones: Their Evolution, Structure and Effects*, Boston, Mass.: American Meteorological Society.

Anthes, R.A. And Trout, J.W. (1971) 'Three-dimensional particle trajectories in a model hurricane', *Weatherwise* 24: 174–8.

Bender, M.A., Tuleya, R.E., and Kurihara, Y. (1987) 'A numerical study of the effect of island terrain on tropical cyclones', *Mon. Wea. Rev.* 115:130–55.

Black, P.G. and Adams, W.L. (1983) *Guidance for Estimating Surface Winds Based on Sea State Observations from Aircraft and Sea State Catalog*, Report from Federal Co-ordinator for Meteorological Services and Supporting Research, Washington, D.C.: FCM–G1–1983, May.

Black, P.G., Burpee, R.W., Dorst, N.M., and Adams, W.L. (1986) 'Appearance of the sea surface in tropical cyclones', *Weather and Forecasting* 1:102–7.

Bosart, L.F. (1984) 'The Texas coastal rainstorm of 17–21 September 1979: an example of synoptic-mesoscale interaction', *Mon. Wea. Rev.* 112: 1108–33.

Brown, L.F., jr., Morton, R.A., McGowen, J.H., Kreitler, C.W., and Fisher, W.L. (1974) *Natural Hazards of the Texas Coastal Zone*, Austin: Bureau of Economic Geology, Box X, Austin, Tex., 78712.

Carter, M.T. (1983) *Probability of Hurricane/Tropical Storm Conditions: A User's Guide for Local Decision Makers*, Washington, DC: National Oceanic and Atmospheric Administration.

Chan, J.C.L. and Gray, W.M. (1982) 'Tropical cyclone movement and surrounding flow relationships', *Mon. Wea. Rev.* 110: 1354–74.

Cooperman, A.I. and Sumner, H.C. (1961) 'North Atlantic tropical cyclones', *Climatological Data, National Summary* 12(9): 468–77.

DeAngelis, R.M. and Hodge, W.T. (1972) *Preliminary Climatic Data Report Hurricane Agnes June 14–23, 1972*, NOAA Technical Memorandum EDS NCC–1, August, Asheville, NC: National Climatic Center.

Dunn, G.E. and Miller, B.I. (1960) *Atlantic Hurricanes*, Baton Rouge: Louisiana State University Press.

Dunn, G.E. and Staff (1967) *Florida Hurricanes*, Fort Worth, Tex.: US Department of Commerce/Environmental Science Services Administration, Technical Memorandum WBTM–SR–38.

Gray, W.M. (1975) *Tropical Cyclone Genesis*, Fort Collins: Colorado State University, Atmospheric Science Paper No. 234.

Gray, W.M. (1981) *Recent Advances in Tropical Cyclone Research From Rawinsonde Composite Analysis* Geneva: World Meteorological Organization.

Gray, W.M. (1989) *Forecast of Atlantic Seasonal Activity for 1989*, Fort Collins: Colorado State University, Atmospheric Science Department.

Holland, G.J. (1983) 'Tropical cyclone motion: environmental interaction plus a beta effect', *J. Atmos. Sci.* 40: 328–42.

Hope, J.F. and Neumann, C.J. (1971) *Digitized Atlantic Tropical Cyclone Tracks*, Fort Worth, Tex.: NOAA Technical Memorandum NWS SR–55.

Jelesnianski, C.P. (1972) *SPLASH (Special Program to List Amplitudes of Surges from Hurricanes, I: Landfall Storms*, Washington, DC: NOAA Technical Memorandum NWS TDL–46, April.

Jelesnianski, C.P. (1974) *SPLASH (Special Program to List Amplitudes of Surges from Hurricanes, II: General Track and Variant Storm Conditions*, Washington, DC: NOAA Technical Memorandum NWS TDL–52, March.

Kotsch, W.J. (1977) *Weather for the Mariner*, 2nd ed., Annapolis, Md.: Naval Institute Press.

Ludlum, D.M. (1963) *Early American Hurricanes, 1492–1870*, Boston, Mass.: American Meteorological Society.

Merrill, R.T. (1985) *Environmental Influences on Hurricane Intensification*, Fort Collins: Colorado State University, Department of Atmospheric Science Paper No. 394.

Merrill, R.T. (1987) *An Experiment in Statistical Prediction of Tropical Cyclone Intensity Change*, Miami, Fla.: NOAA Technical Memorandum NWS NHC 34.

Neumann, C.J. Jarvinen, B.R., Pike, A.C. (1987) *Tropical Cyclones of the North Atlantic Ocean, 1871–1986*. Coral Gables, Fla.: National Climatic Data Center in cooperation with the National Hurricane Center. (For sale by the National Environmental Satellite, Data, and Information Service, National Climatic Data Center, Asheville, NC 28801. Maps for 1987–89 have been added.)

Novlan, D.J. and Gray, W.M. (1974) 'Hurricane-spawned tornadoes', *Mon. Wea. Rev.* 102: 476–88.

Orton, R.B. (1970) 'Tornadoes associated with Hurricane Beulah on September 19–23, 1967', *Mon. Wea. Rev.* 98: 541–47.

Orton, R.B. and Condon, C.R. (1970) 'Hurricane Celia, July 30–August 5', *Climatological Data, National Summary* 21: 403–18.

Sheets, R.C. (1981) *Tropical Cyclone Modification: The Project Stormfury Hypothesis*, Miami, Fla: NOAA Technical Report ERL 414–AOML30, 52 pg.

Simpson, R.H. (1954) 'Hurricanes', *Scientific American*, June.

Simpson, R.H. (1970) 'The Atlantic hurricane season of 1969', *Mon. Wea. Rev.* 98: 293–306.

Simpson, R.H. (1971) *The Decision Process in Hurricane Forecasting*, Fort Worth, Tex.: NOAA Technical Memorandum NWS SR–53.

Simpson, R.H. and Riehl, H. (1981) *The Hurricane and Its Impact*, Baton Rouge: Louisiana State University Press.

Simpson, R.H. *et al.* (1978) *TYMOD: Typhoon Moderation*, Final Report prepared for the Government of the Philippines, Arlington: Virginia Technology.

Texas Coastal and Marine Council (1974) *Hurricane Awareness Briefings*, Austin: Texas Coastal and Marine Council.

Texas Coastal and Marine Council (1976) *Model Minimum Hurricane-Resistant Building Standards for the Texas Gulf Coast*, Austin: Texas Coastal and Marine Council.

US Army Corps of Engineers (1962) *Report on Hurricane Carla, 9–12 September 1961*, Galveston, Tex.: US Army Corps of Engineers, Galveston District.

US Department of Commerce (1964) *Hurricane Dora, August 28–September 16, 1964*, Preliminary Report with Advisories and Bulletins Issued, Washington, DC.

US Department of Commerce (1965) *Hurricane Betsy, August 27–September 12, 1965*, Preliminary Report with Advisories and Bulletins Issued, Washington, DC.

US Department of Commerce (1969) *The Virginia Floods, August 19–22, 1969*, A Report to the Administrator, September, Washington, DC.

US Department of Commerce (1973) *National Disaster Survey Report 73–1, Final Report of the Disaster Survey Team on the Advents of Agnes*, A Report to the Administrator, Rockville, Md.

US Department of Commerce (1977) *Hurricanes, Florida and You*, Washington, DC: National Oceanic and Atmospheric Administration, National Weather Service.

White, Glenn, G. (1982) *The Global Circulation of the Atmosphere, December 1980–November 1981, based on ECMWF Analyses*. Reading, Berks: University of Reading.

Willoughby, H.E. (1979) *Some Aspects of the Dynamics in Hurricane Anita of 1977*, Coral Gables, Fla: NOAA Technical Memorandum, ERL NHEML–5.

Further reading

Anthes, R.A. (1982) *Tropical Cyclones: Their Evolution, Structure and Effects*, Boston, Mass.: American Meteorological Society.

This book provides an excellent post-graduate-level discussion of tropical cyclones. It is particularly valuable for readers with prior knowledge of meteorology.

Burpee, R.W. (1988) 'Grady Norton: Hurricane forecaster and communicator extraordinaire', *Weather and Forecasting* 3: 247–54.

This paper presents a biography of Atlantic hurricane forecaster Grady Norton, including perspectives from fellow hurricane forecasters and researchers, Gordon Dunn, Cecil Gentry, Paul Moore, and Bob Simpson.

Burpee, R.W. (1989) 'Gordon E. Dunn: preeminent forecaster of midlatitude storms and tropical cyclones', *Weather and Forecasting* 4: 573–84.

This article presents a biography of tropical cyclone forecaster Gordon E. Dunn, including reminiscences by Cecil Gentry, Len Snellman, Arnold Sugg, and Gilbert Clark.

Dunn, G.E. and Miller, B.I. (1960) *Atlantic Hurricanes*, Baton Rouge: Louisiana State University Press.

This book, written about 30 years ago, remains a valuable source in the description of Atlantic tropical cyclones. Although some of the material requires knowledge of statistics, the educated lay reader will find considerable useful material in this source.

Elsberry, R.L. (ed.) (1988) *A Global View of Tropical Cyclones, International Workshop on Tropical Cyclones, Bangkok, Thailand*, Monterey, Calif.: Naval Postgraduate School.

This book is a recent collection of papers from an international workshop on tropical cyclones held in Bangkok, Thailand from 25 November–5 December 1985. The topics include statements on the current state of tropical cyclone research, forecasting, and warning procedures.

Ludlum, D.M. (1963) *Early American Hurricanes, 1492–1870*, Boston, Mass.: American Meteorological Society.

D.M. Ludlum has completed a number of informative books concerning early American weather, and this volume summarizes existing knowledge about tropical cyclones in eastern North America between 1492 and 1870. A new printing of this book is to be published in 1990.

Simpson, R.H. and Riehl, H. (1981) *The Hurricane and Its Impact*, Baton Rouge: Louisiana State University Press.

This text, although designed for engineers who require knowledge concerning tropical cyclones, is also an extremely informative source for the educated lay public. It includes a step-by-step description of a hypothetical hurricane landfall on the west coast of Florida.

Index

active front
  defined 22
Atlantic tropical cyclone tracking map
  222
Atlantic tropical cyclone tracks 9,
  15–17, 101–219
autumnal equinox
  defined 16

baroclinicity
  defined 21–2
Beaufort number
  defined in terms of wind speed
  68–70
bulletins
  current forecasts 96
  Hurricane Camille 1, 2

centrifugal force
  defined 35
centripetal acceleration
  defined 35
circumpolar vortex
  defined 22
coastal zoning criteria 78–81
cold fronts
  defined 22
conditions required to maintain
  hurricane intensity 41–4
controls on tropical cyclone
  movement
  external controls 48–50
  influence of terrain 54–5
  interaction of the external controls
    and the tropical cyclone 50–5
  internal influences 55–7
convergence influence on
  cumulonimbus development 33
Coriolis effect
  defined 20
  discussed 34
  influence on tropical cyclone
    movement 53
cyclogenesis
  defined 22

direct heat engine
  Hadley cell 21
  tropical cyclone 33

El Niño
  defined 97

used in seasonal tropical cyclone
  forecasting 96–7
equatorial trough
  defined 20
European Centre for Medium-range
  Weather Forecasts (ECMWF)
  global model 90
explosive cyclogenesis
  defined 22
extratropical cyclone
  defined 21–2
  preferred location for
    development 23
  relation to tropical cyclone 25
eyes
  defined 35
  discussed 35–40
  radar depiction of
    discussed 35–6
    illustrated 37–40
  relation to tropical cyclone
    modification 97, 99–100
eyewall
  defined 35
  double eyewall 36

feeder bands
  defined 35
flash flood warning
  defined 96
flash flood watch
  defined 94
flood forecast bulletin
  defined 96
forecast decision trees 94–5

gale warning
  defined 91
general circulation of the earth 11–12,
  20–5
Hadley cell
  defined 21
  influenced by circular high
    pressure systems 24
  relation to tropical cyclone
    development 25
  relation to tropical cyclone
    movement 25
HURRAN model 89–90
hurricane
  defined 3
Hurricane Camille 1–3

hurricane warning
  defined 91
hurricane watch
  defined 91

influence of mountains on weather
  patterns 25
intertropical convergence zone
  defined 21
  discussed 25, 32, 34
  illustrated 23

jet stream
  defined 22

latent heat
  defined 33

major criteria for the development of
  tropical cyclones
  favourable geographic locations as
    related to development criteria 46
  summarized 44–5
major criteria for the weakening of
  tropical cyclones 45–6
maximum wind
  relation to central pressure 41
  relation to sea surface pressure
    42–3
  sustained wind estimate in a
    tropical cyclone 37
Moveable Fine-mesh Model 90
Multiple–Nested Grid Model 90

names of tropical cyclones 88
Nested Tropical Cyclone Model 90
numerical modelling
  simulation of air motion in a
    tropical cyclone 39
  simulation of the influence of
    terrain on tropical cyclone
    motion 55
  simulation of storm surge 71, 74
  use in forecasting tropical cyclone
    track 89–90

occluded front
  defined 22
ocean surface
  aerodynamic roughness of the
    surface
    formula for 73

observed in tropical cyclones 60–8
verbal description in tropical
    cyclones 68–70
One-way Tropical Cyclone Model 90
outflow jets
    defined 53
    influence on tropical cyclone
        movement 53–4

polar front
    defined 20
probability of landfall
    public advisory 93
Project Stormfury 99–100
public forecasts of tropical cyclone
    associated weather 91–4, 96

quasi-biennial oscillation
    defined 23

radar
    image of the eye of a tropical
        cyclone 40
    movement of a tropical cylone
        51–2
    track of the eye of a tropical
        cyclone 56
radius of maximum winds
    defined 35
    discussed 38–9
    illustrated 44
rainfall associated with tropical
    cyclones 75–7, 83–7
references 223–6

Saffir/Simpson Damage-Potential
    Scale
    defined 4
    discussed 37
    example of its potential use 10
    example of use 19
SANBAR model 89–90
satellite
    infra-red image of a tropical
        cyclone 36
    visible image of a tropical cyclone
        38
seasonal prediction of tropical cyclone
    activity 96–8
severe thunderstorm warning 94
severe thunderstorm watch 94
small craft cautionary statements 91
snow associated with tropical cyclones
    82
stationary front
    defined 22
statistical models
    NHC 67, NHC 72, NHC 73 89
storm surge
    defined 59
    discussed 59, 71, 74
storm warning
    defined 91

stratosphere
    defined 12
    discussed 32
subtropical ridge
    defined 21
    influence on tropical cyclone
        movement 48–50
summer monsoon
    defined 24
supercooled water
    defined 99
sustained wind in a tropical cyclone
    distribution as a function of
        distance from the centre 44, 72
    distribution with height over land
        and water 45
    relation between wind speed and
        kinetic energy 72, 75

temperature in a tropical cyclone
    distribution with height over land
        and water 45
thermocline
    defined 24
tornado warning
    defined 94
tornado watch
    defined 93
tornadoes associated with tropical
    cylones
    discussed 77–9
    hypothesized mechanism for
        development 77
trade winds
    defined 21
tropical cyclone
    controls on tropical cyclone
        movement 48–57
    defined 3
    developed criteria 25–6
    forecasts of 88–100
    formation 32–4
    further reading 226
    genesis location 5–8
    geographical and seasonal
        distribution 5–31
    impacts of 58–87
    intensification 34–47
    maps which can be used to
        determine geographic locations
        of favourable conditions for
        development 26–31
    mechanisms of formation and
        development 32–47
    modification of 97, 99–100
    movement 8–17
    probability of being observed in
        the Atlantic, Gulf of Mexico,
        and Caribbean as a function of
        day of the year 19
    probability of occurrence in the
        Atlantic, Gulf of Mexico, and
        Caribbean within 2.5 degree
        latitude–longitude areas 18

track forecasts 89–93
tracks 9–17, 49, 51, 52, 55, 56,
    73, 79, 89, 102–219, 222
tropical cyclone modification
    discussed 97, 99–100
    illustration of modification
        hypothesis 99
tropical depression
    defined 3
tropical disturbance
    defined 3
tropical low
    defined 3
tropical storm
    defined 3
tropopause
    defined 12
    value of in the equatorial trough
        20
typhoon
    defined 3

upwelling
    defined 24

vernal equinox
    defined 10

warm front
    defined 22
wave cyclones
    defined 22
waves
    difference in tropical cyclones and
        extratropical cyclones 58–9
    influenced by fetch 70
    observed in tropical cyclones
        68–70
weather forecasting
    available by telephone 96
    decision tree for tropical cyclone
        development 94
    decision tree for tropical cyclone
        intensification 95
    public forecasts of tropical cyclone
        related weather 1–2, 91–4, 96
    of rainfall associated with the
        remnants of Camille (1969) 83
    relation to general circulation
        patterns 25
    seasonal forecasting of tropical
        cyclone activity 96–7
    of tropical cyclone tracks 89–93
winds in a tropical cyclone
    distribution as a function of
        distance from the centre 44, 72
    distribution with height over land
        and water 45
    relation to maximum wind gusts
        72–3
    relation between wind speed and
        kinetic energy 72, 75
winter monsoon
    defined 24

T - #0227 - 071024 - C0 - 246/174/13 - PB - 9780415615549 - Gloss Lamination